The Wonders of Instinct

The Wonders of Instinct

J. H. Fabre

THE WONDERS OF INSTINCT

Published in the United States by IndyPublish.com
McLean, Virginia

ISBN 1-4043-3900-0 (hardcover)
ISBN 1-4043-3901-9 (paperback)

CONTENTS.

INDEX.

Note:—Chapters 5 and 6 have been translated by Mr. Bernard Miall; the remainder by Mr. Alexander Teixeira de Mattos.

ILLUSTRATIONS.

EXPERIMENT 2. A dead mouse is placed on the branches of a tuft of thyme. By dint of jerking, shaking and tugging at the body, the Burying-beetles succeed in extricating it from the twigs and bringing it down.

EXPERIMENT 3. With a ligament of raphia, the Mole is fixed by the hind feet to a twig planted vertically in the soil. The head and shoulders touch the ground. By digging under these, the Necrophori at the same time uproot the gibbet, which eventually falls, dragged over by the weight of its burden.

EXPERIMENT 4. The stake is slanting; the Mole touches the ground, but at a point two inches from the base of the gibbet. The Burying-beetles begin by digging to no purpose under the body. They make no attempt to overturn the stake. In this experiment they obtain the Mole at last by employing the usual method, that is by gnawing the bond.

THE BLUEBOTTLE LAYING HER EGGS IN THE SLIT OF A DEAD BIRD'S BEAK.

THE LYCOSA LIFTING HER WHITE BAG OF EGGS TOWARDS THE SUN, TO ASSIST THE HATCHING.
The Lycosa lying head downwards on the edge of her pit, holding in her hind-legs her white bag of eggs and lifting them towards the sun, to assist the hatching.

THE BANDED EPEIRA INSCRIBING HER FLOURISH, AFTER FINISHING HER WEB.

THE BANDED EPEIRA LETTING HERSELF DROP BY THE END OF HER THREAD.

THE BANDED EPEIRA SWATHING HER CAPTURE.
The web has given way in many places during the struggle.

OSMIA-NESTS IN A BRAMBLE TWIG.

OSMIA-NESTS INSIDE A REED.

ARTIFICIAL HIVE INVENTED BY THE AUTHOR TO STUDY THE OSMIA'S LAYING.
It consists of reed-stumps arranged Pan-pipe fashion.

OLD NESTS USED BY THE OSMIA IN LAYING HER EGGS.

1. Nest of the Mason-bee of the Shrubs.

2. Osmia-grubs in empty shells of the Garden Snail.

3. Nests of the Mason-bee of the Sheds.

THE GLOW-WORM: a, male; b, female.

THE CABBAGE CATERPILLAR: a, the caterpillars; b, the cocoons of their parasite, Microgaster glomeratus.

CHAPTER 1.

THE HARMAS.

This is what I wished for, hoc erat in votis: a bit of land, oh, not so very large, but fenced in, to avoid the drawbacks of a public way; an abandoned, barren, sun-scorched bit of land, favoured by thistles and by Wasps and Bees. Here, without fear of being troubled by the passers-by, I could consult the Ammophila and the Sphex (two species of Digger-or Hunting-wasps.—Translator's Note.) and engage in that difficult conversation whose questions and answers have experiment for their language; here, without distant expeditions that take up my time, without tiring rambles that strain my nerves, I could contrive my plans of attack, lay my ambushes and watch their effects at every hour of the day. Hoc erat in votis. Yes, this was my wish, my dream, always cherished, always vanishing into the mists of the future.

And it is no easy matter to acquire a laboratory in the open fields, when harassed by a terrible anxiety about one's daily bread. For forty years have I fought, with steadfast courage, against the paltry plagues of life; and the long-wished-for labo-ratory has come at last. What it has cost me in perseverance and relentless work I will not try to say. It has come; and, with it—a more serious condition—perhaps a little leisure. I say perhaps, for my leg is still hampered with a few links of the convict's chain.

The wish is realized. It is a little late, O! my pretty insects! I greatly fear that the peach is offered to me when I am beginning to have no teeth wherewith to eat it. Yes, it is a little late: the wide horizons of the outset have shrunk into a low and stifling canopy, more and more straitened day by day. Regretting nothing in the past, save those whom I have lost; regretting nothing, not even my first youth; hoping nothing either, I have reached the point at which, worn out by the experience of things, we ask ourselves if life be worth the living.

Amid the ruins that surround me, one strip of wall remains standing, immovable upon its solid base: my passion for scientific truth. Is that enough, O! my busy insects, to enable me to add yet a few seemly pages to your history? Will my strength not cheat my good intentions? Why, indeed, did I forsake you so long?

Friends have reproached me for it. Ah, tell them, tell those friends, who are yours as well as mine, tell them that it was not forgetfulness on my part, not weariness, nor neglect: I thought of you; I was convinced that the Cerceris' (A species of Digger-wasp.—Translator's Note.) cave had more fair secrets to reveal to us, that the chase of the Sphex held fresh surprises in store. But time failed me; I was alone, deserted, struggling against misfortune. Before philosophizing, one had to live. Tell them that, and they will pardon me.

Others have reproached me with my style, which has not the solemnity, nay, better, the dryness of the schools. They fear lest a page that is read without fatigue should not always be the expression of the truth. Were I to take their word for it, we are profound only on condition of being obscure. Come here, one and all of you—you, the sting-bearers, and you, the wing-cased armour-clads—take up my defence and bear witness in my favour. Tell of the intimate terms on which I live with you, of the patience with which I observe you, of the care with which I record your actions. Your evidence is unanimous: yes, my pages, though they bristle not with hollow formulas nor learned smatterings, are the exact narrative of facts observed, neither more nor less; and whoso cares to question you in his turn will obtain the same replies.

And then, my dear insects, if you cannot convince those good people, because you do not carry the weight of tedium, I, in my turn, will say to them:

"You rip up the animal and I study it alive; you turn it into an object of horror and pity, whereas I cause it to be loved; you labour in a torture-chamber and dissecting-room, I make my observations under the blue sky, to the song of the Cicadae (The Cicada Cigale, an insect akin to the Grasshopper and found more particularly in the south of France.—Translator's Note.); you subject cell and protoplasm to chemical tests, I study instinct in its loftiest manifestations; you pry into death, I pry into life. And why should I not complete my thought: the boars have muddied the clear stream; natural history, youth's glorious study, has, by dint of cellular improvements, become a hateful and repulsive thing. Well, if I write for men of learning, for philosophers, who, one day, will try to some extent to unravel the tough problem of instinct, I write also, I write above all things, for the young, I want to make them love the natural history which you make them hate; and that is why, while keeping strictly to the domain of truth, I avoid your scientific prose, which too often, alas, seems borrowed from some Iroquois idiom!"

But this is not my business for the moment: I want to speak of the bit of land long cherished in my plans to form a laboratory of living entomology, the bit of land which I have at last obtained in the solitude of a little village. It is a "harmas," the name given, in this district (The country round Sérignan, in Provence.—Translator's Note.), to an untilled, pebbly expanse abandoned to the vegetation of the thyme. It is too poor to repay the work of the plough; but the Sheep passes there in spring, when it has chanced to rain and a little grass shoots up.

My harmas, however, because of its modicum of red earth swamped by a huge mass of stones, has received a rough first attempt at cultivation: I am told that vines once grew here. And, in fact, when we dig the ground before planting a few trees, we turn up, here and there, remains of the precious stock, half carbonized by time. The three-pronged fork, therefore, the only implement of husbandry that can penetrate such a soil as this, has entered here; and I am sorry, for the primitive vegetation has disappeared. No more thyme, no more lavender, no more clumps of kermes-oak, the dwarf oak that forms forests across which we step by lengthening our stride a little. As these plants, especially the first two, might be of use to me by offering the Bees and Wasps a spoil to forage, I am compelled to reinstate them in the ground whence they were driven by the fork.

What abounds without my mediation is the invaders of any soil that is first dug up and then left for a time to its own resources. We have, in the first rank, the couch-grass, that execrable weed which three years of stubborn warfare have not succeeded in exterminating. Next, in respect of number, come the centauries, grim-looking one and all, bristling with prickles or starry halberds. They are the yellow-flowered centaury, the mountain centaury, the star-thistle and the rough centaury: the first predominates. Here and there, amid their inextricable confusion, stands, like a chandelier with spreading orange flowers for lights, the fierce Spanish oyster-plant, whose spikes are strong as nails. Above it towers the Illyrian cotton-thistle, whose straight and solitary stalk soars to a height of three to six feet and ends in large pink tufts. Its armour hardly yields before that of the oyster-plant. Nor must we forget the lesser thistle-tribe, with, first of all, the prickly or "cruel" thistle, which is so well armed that the plant-collector knows not where to grasp it; next, the spear-thistle, with its ample foliage, ending each of its veins with a spear-head; lastly, the black knap-weed, which gathers itself into a spiky knot. In among these, in long lines armed with hooks, the shoots of the blue dewberry creep along the ground. To visit the prickly thicket when the Wasp goes foraging, you must wear boots that come to mid-leg or else resign yourself to a smarting in the calves. As long as the ground retains a few remnants of the vernal rains, this rude vegetation does not lack a certain charm, when the pyramids of the oyster-plant and the slender branches of the cotton-thistle rise above the wide carpet formed by the yellow-flowered centaury's saffron heads; but let the droughts of summer come and we see but a desolate waste, which the flame of a match would set ablaze from one end to the other. Such is, or rather was, when I took possession of it, the Eden of bliss where I mean to live henceforth alone with the insect. Forty years of desperate struggle have won it for me.

Eden, I said; and, from the point of view that interests me, the expression is not out of place. This cursed ground, which no one would have had at a gift to sow with a pinch of turnip-seed, is an earthly paradise for the Bees and the Wasps. Its mighty growth of thistles and centauries draws them all to me from everywhere around. Never, in my insect-hunting memories, have I seen so large a population at a single spot; all the trades have made it their rallying-point. Here come hunters of every kind of game, builders in clay, weavers of cotton goods, collectors of pieces cut from a leaf or the petals of a flower, architects in paste-board, plasterers mixing mortar, carpenters boring wood, miners digging underground galleries, workers handling goldbeater's skin and many more.

Who is this one? An Anthidium. (A Cotton-bee.—Translator's Note.) She scrapes the cobwebby stalk of the yellow-flowered centaury and gathers a ball of wadding which she carries off proudly in the tips of her mandibles. She will turn it, under ground, into cotton-felt satchels to hold the store of honey and the egg. And these others, so eager for plunder? They are Megachiles (Leaf-cutting Bees.—Translator's Note.), carrying under their bellies their black, white, or blood-red reaping-brushes. They will leave the thistles to visit the neighbouring shrubs and there cut from the leaves oval pieces which will be made into a fit receptacle to contain the harvest. And these, clad in black velvet? They are Chalicodomae (Mason-bees.—Translator's Note.), who work with cement and gravel. We could easily find their masonry on the stones in the harmas. And these, noisily buzzing with a sudden flight? They are the Anthophorae (a species of Wild Bees.—Translator's Note.), who live in the old walls and the sunny banks of the neighbourhood.

Now come the Osmiae. One stacks her cells in the spiral staircase of an empty snail-shell; another, attacking the pith of a dry bit of bramble, obtains for her

grubs a cylindrical lodging and divides it into floors by means of partition-walls;
a third employs the natural channel of a cut reed; a fourth is a rent-free tenant of
the vacant galleries of some Mason-bee. Here are the Macrocerae and the Eucerae,
whose males are proudly horned; the Dasypodae, who carry an ample brush of
bristles on their hind-legs for a reaping implement; the Andrenae, so manyfold in
species; the slender-bellied Halicti. (Osmiae, Macrocerae, Eucerae, Dasypodae,
Andrenae, and Halicti are all different species of Wild Bees.—Translator's Note.)
I omit a host of others. If I tried to continue this record of the guests of my thistles, it would muster almost the whole of the honey-yielding tribe. A learned
entomologist of Bordeaux, Professor P,rez, to whom I submit the naming of my
prizes, once asked me if I had any special means of hunting, to send him so many
rarities and even novelties. I am not at all an experienced and still less a zealous
hunter, for the insect interests me much more when engaged in its work than
when stuck on a pin in a cabinet. The whole secret of my hunting is reduced to
my dense nursery of thistles and centauries.

By a most fortunate chance, with this populous family of honey-gatherers was
allied the whole hunting tribe. The builders' men had distributed here and there,
in the harmas, great mounds of sand and heaps of stones, with a view of running
up some surrounding walls. The work dragged on slowly; and the materials found
occupants from the first year. The Mason-bees had chosen the interstices between
the stones as a dormitory where to pass the night in serried groups. The powerful
Eyed Lizard, who, when close-pressed, attacks wide-mouthed both man and dog,
had selected a cave wherein to lie in wait for the passing Scarab (A Dung-beetle
known also as the Sacred Beetle.—Translator's Note.); the Black-eared Chat,
garbed like a Dominican, white-frocked with black wings, sat on the top stone,
singing his short rustic lay: his nest, with its sky-blue eggs, must be somewhere in
the heap. The little Dominican disappeared with the loads of stones. I regret him:
he would have been a charming neighbour. The Eyed Lizard I do not regret at all.

The sand sheltered a different colony. Here, the Bembeces (A species of Digger-
wasps.—Translator's Note.) were sweeping the threshold of their burrows, flinging a curve of dust behind them; the Languedocian Sphex was dragging her
Ephippigera (A species of Green Grasshopper—Translator's Note.) by the antennae; a Stizus (A species of Hunting-wasp.—Translator's Note.) was storing her
preserves of Cicadellae. (Froghoppers—Translator's Note.) To my sorrow, the
masons ended by evicting the sporting tribe; but, should I ever wish to recall it, I
have but to renew the mounds of sand: they will soon all be there.

Hunters that have not disappeared, their homes being different, are the
Ammophilae, whom I see fluttering, one in spring, the others in autumn, along

the garden-walks and over the lawns, in search of a caterpillar; the Pompili (The Pompilus is a species of Hunting-wasp known also as the Ringed Calicurgus—Translator's Note.), who travel alertly, beating their wings and rummaging in every corner in quest of a Spider. The largest of them waylays the Narbonne Lycosa (Known also as the Black-bellied Tarantula—Translator's Note.), whose burrow is not infrequent in the harmas. This burrow is a vertical well, with a curb of fescue-grass intertwined with silk. You can see the eyes of the mighty Spider gleam at the bottom of the den like little diamonds, an object of terror to most. What a prey and what dangerous hunting for the Pompilus! And here, on a hot summer afternoon, is the Amazon-ant, who leaves her barrack-rooms in long battalions and marches far afield to hunt for slaves. We will follow her in her raids when we find time. Here again, around a heap of grasses turned to mould, are Scoliae (Large Hunting-wasps—Translator's Note.) an inch and a half long, who fly gracefully and dive into the heap, attracted by a rich prey, the grubs of Lamellicorns, Oryctes, and Cetoniae. (Different species of Beetles. The Cetonia is the Rose-chafer—Translator's Note.)

What subjects for study! And there are more to come. The house was as utterly deserted as the ground. When man was gone and peace assured, the animal hastily seized on everything. The Warbler took up his abode in the lilac-shrubs; the Greenfinch settled in the thick shelter of the cypresses; the Sparrow carted rags and straw under every slate; the Serin-finch, whose downy nest is no bigger than half an apricot, came and chirped in the plane-tree tops; the Scops made a habit of uttering his monotonous, piping note here, of an evening; the bird of Pallas Athene, the Owl, came hurrying along to hoot and hiss.

In front of the house is a large pond, fed by the aqueduct that supplies the village pumps with water. Here, from half a mile and more around, come the Frogs and Toads in the lovers' season. The Natterjack, sometimes as large as a plate, with a narrow stripe of yellow down his back, makes his appointments here to take his bath; when the evening twilight falls, we see hopping along the edge the Midwife Toad, the male, who carries a cluster of eggs, the size of peppercorns, wrapped round his hind-legs: the genial paterfamilias has brought his precious packet from afar, to leave it in the water and afterwards retire under some flat stone, whence he will emit a sound like a tinkling bell. Lastly, when not croaking amid the foliage, the Tree-frogs indulge in the most graceful dives. And so, in May, as soon as it is dark, the pond becomes a deafening orchestra: it is impossible to talk at table, impossible to sleep. We had to remedy this by means perhaps a little too rigorous. What could we do? He who tries to sleep and cannot needs become ruthless.

Bolder still, the Wasp has taken possession of the dwelling-house. On my door-sill, in a soil of rubbish, nestles the White-banded Sphex: when I go indoors, I must be careful not to damage her burrows, not to tread upon the miner absorbed in her work. It is quite a quarter of a century since I last saw the saucy Cricket-hunter. When I made her acquaintance, I used to visit her at a few miles' distance: each time, it meant an expedition under the blazing August sun. To-day I find her at my door; we are intimate neighbours. The embrasure of the closed window provides an apartment of a mild temperature for the Pelopaeus. (A species of Mason-wasp—Translator's Note.) The earth-built nest is fixed against the free-stone wall. To enter her home, the Spider-huntress uses a little hole left open by accident in the shutters. On the mouldings of the Venetian blinds, a few stray Mason-bees build their group of cells; inside the outer shutters, left ajar, a Eumenes (Another Mason-wasp—Translator's Note.) constructs her little earthen dome, surmounted by a short, bell-mouthed neck. The Common Wasp and the Polistes (A Wasp that builds her nest in trees—Translator's Note.) are my dinner-guests: they visit my table to see if the grapes served are as ripe as they look.

Here surely—and the list is far from complete—is a company both numerous and select, whose conversation will not fail to charm my solitude, if I succeed in draw-ing it out. my dear beasts of former days, my old friends, and others, more recent acquaintances, all are here, hunting, foraging, building in close proximity. Besides, should we wish to vary the scene of observation, the mountain (Mont Ventoux, an outlying summit of the Alps, 6,270 feet high.—Translator's Note.) is but a few hundred steps away, with its tangle of arbutus, rock-roses and arbores-cent heather; with its sandy spaces dear to the Bembeces; with its marly slopes exploited by different Wasps and Bees. And that is why, foreseeing these riches, I have abandoned the town for the village and come to Sérignan to weed my turnips and water my lettuces.

Laboratories are being founded at great expense, on our Atlantic and Mediterranean coasts, where people cut up small sea-animals, of but meagre inter-est to us; they spend a fortune on powerful microscopes, delicate dissecting-instruments, engines of capture, boats, fishing-crews, aquariums, to find out how the yolk of an Annelid's (A red-blooded Worm.—Translator's Note.) egg is con-structed, a question whereof I have never yet been able to grasp the full impor-tance; and they scorn the little land-animal, which lives in constant touch with us, which provides universal psychology with documents of inestimable value, which too often threatens the public wealth by destroying our crops. When shall we have an entomological laboratory for the study not of the dead insect, steeped in alcohol, but of the living insect; a laboratory having for its object the instinct, the habits, the manner of living, the work, the struggles, the propagation of that little world with which agriculture and philosophy have most seriously to reckon? To know thoroughly the history of the destroyer of our vines might perhaps be more important than to know how this or that nerve-fibre of a Cirriped ends (Cirripeds are sea-animals with hair-like legs, including the Barnacles and Acorn-shells.—Translator's Note.); to establish by experiment the line of demarcation between intellect and instinct; to prove, by comparing facts in the zoological pro-gression, whether human reason be an irreducible faculty or not: all this ought surely to take precedence of the number of joints in a Crustacean's antenna. These enormous questions would need an army of workers; and we have not one. The fashion is all for the Mollusc and the Zoophyte. (Zoophytes are plant-like sea-ani-mals, including Star-fishes, Jelly-fishes, Sea-anemones, and Sponges.—Translator's Note.) The depths of the sea are explored with many drag-nets; the soil which we tread is consistently disregarded. While waiting for the fashion to change, I open my harmas laboratory of living entomology; and this laboratory shall not cost the ratepayers one farthing.

CHAPTER 2.

THE GREEN GRASSHOPPER.

We are in the middle of July. The astronomical dog-days are just beginning; but in reality the torrid season has anticipated the calendar and for some weeks past the heat has been overpowering.

This evening in the village they are celebrating the National Festival. (The 14th of July, the anniversary of the fall of the Bastille.—Translator's Note.) While the little boys and girls are hopping round a bonfire whose gleams are reflected upon the church-steeple, while the drum is pounded to mark the ascent of each rocket, I am sitting alone in a dark corner, in the comparative coolness that prevails at nine o'clock, harking to the concert of the festival of the fields, the festival of the harvest, grander by far than that which, at this moment, is being celebrated in the village square with gunpowder, lighted torches, Chinese lanterns and, above all, strong drink. It has the simplicity of beauty and the repose of strength.

It is late; and the Cicadae are silent. Glutted with light and heat, they have indulged in symphonies all the livelong day. The advent of the night means rest for them, but a rest frequently disturbed. In the dense branches of the plane-trees a sudden sound rings out like a cry of anguish, strident and short. It is the desperate wail of the Cicada, surprised in his quietude by the Green Grasshopper, that ardent nocturnal huntress, who springs upon him, grips him in the side, opens and ransacks his abdomen. An orgy of music, followed by butchery.

I have never seen and never shall see that supreme expression of our national revelry, the military review at Longchamp; nor do I much regret it. The newspapers tell me as much about it as I want to know. They give me a sketch of the site. I see, installed here and there amid the trees, the ominous Red Cross, with the legend, "Military Ambulance; Civil Ambulance." There will be bones broken, apparently; cases of sunstroke; regrettable deaths, perhaps. It is all provided for and all in the programme.

Even here, in my village, usually so peaceable, the festival will not end, I am ready to wager, without the exchange of a few blows, that compulsory seasoning of a day of merry-making. No pleasure, it appears, can be fully relished without an added condiment of pain.

Let us listen and meditate far from the tumult. While the disembowelled Cicada utters his protest, the festival up there in the plane-trees is continued with a change of orchestra. It is now the time of the nocturnal performers. Hard by the place of slaughter, in the green bushes, a delicate ear perceives the hum of the Grasshoppers. It is the sort of noise that a spinning-wheel makes, a very unobtrusive sound, a vague rustle of dry membranes rubbed together. Above this dull bass there rises, at intervals, a hurried, very shrill, almost metallic clicking. There you have the air and the recitative, intersected by pauses. The rest is the accompaniment.

Despite the assistance of a bass, it is a poor concert, very poor indeed, though there are about ten executants in my immediate vicinity. The tone lacks intensity. My old tympanum is not always capable of perceiving these subtleties of sound. The little that reaches me is extremely sweet and most appropriate to the calm of twilight. Just a little more breadth in your bow-stroke, my dear Green Grasshopper, and your technique would be better than the hoarse Cicada's, whose name and reputation you have been made to usurp in the countries of the north.

Still, you will never equal your neighbour, the little bell-ringing Toad, who goes tinkling all round, at the foot of the plane-trees, while you click up above. He is the smallest of my batrachian folk and the most venturesome in his expeditions.

How often, at nightfall, by the last glimmers of daylight, have I not come upon him as I wandered through my garden, hunting for ideas! Something runs away, rolling over and over in front of me. Is it a dead leaf blown along by the wind? No, it is the pretty little Toad disturbed in the midst of his pilgrimage. He hurriedly takes shelter under a stone, a clod of earth, a tuft of grass, recovers from his excitement and loses no time in picking up his liquid note.

On this evening of national rejoicing, there are nearly a dozen of him tinkling against one another around me. Most of them are crouching among the rows of flower-pots that form a sort of lobby outside my house. Each has his own note, always the same, lower in one case, higher in another, a short, clear note, melodious and of exquisite purity.

With their slow, rhythmical cadence, they seem to be intoning litanies. "Cluck," says one; "click," responds another, on a finer note; "clock," adds a third, the tenor of the band. And this is repeated indefinitely, like the bells of the village pealing on a holiday: "cluck, click, clock; cluck, click, clock!"

The batrachian choristers remind me of a certain harmonica which I used to covet when my six-year-old ear began to awaken to the magic of sounds. It consisted of a series of strips of glass of unequal length, hung on two stretched tapes. A cork fixed to a wire served as a hammer. Imagine an unskilled hand striking at random on this key-board, with a sudden clash of octaves, dissonances and topsy-turvy chords; and you will have a pretty clear idea of the Toads' litany.

As a song, this litany has neither head nor tail to it; as a collection of pure sounds, it is delicious. This is the case with all the music in nature's concerts. Our ear discovers superb notes in it and then becomes refined and acquires, outside the realities of sound, that sense of order which is the first condition of beauty.

Now this sweet ringing of bells between hiding-place and hiding-place is the matrimonial oratorio, the discreet summons which every Jack issues to his Jill. The sequel to the concert may be guessed without further enquiry; but what it would be impossible to foresee is the strange finale of the wedding. Behold the father, in this case a real paterfamilias, in the noblest sense of the word, coming out of his retreat one day in an unrecognizable state. He is carrying the future, tight-packed around his hind-legs; he is changing houses laden with a cluster of eggs the size of peppercorns. His calves are girt, his thighs are sheathed with the bulky burden; and it covers his back like a beggar's wallet, completely deforming him.

Whither is he going, dragging himself along, incapable of jumping, thanks to the weight of his load? He is going, the fond parent, where the mother refuses to go; he is on his way to the nearest pond, whose warm waters are indispensable to the tadpoles' hatching and existence. When the eggs are nicely ripened around his legs under the humid shelter of a stone, he braves the damp and the daylight, he the passionate lover of dry land and darkness; he advances by short stages, his lungs congested with fatigue. The pond is far away, perhaps; no matter: the plucky pilgrim will find it.

He's there. Without delay, he dives, despite his profound antipathy to bathing; and the cluster of eggs is instantly removed by the legs rubbing against each other.

The eggs are now in their element; and the rest will be accomplished of itself. Having fulfilled his obligation to go right under, the father hastens to return to his well-sheltered home. He is scarcely out of sight before the little black tadpoles are hatched and playing about. They were but waiting for the contact of the water in order to burst their shells.

Among the singers in the July gloaming, one alone, were he able to vary his notes, could vie with the Toad's harmonious bells. This is the little Scops-owl, that comely nocturnal bird of prey, with the round gold eyes. He sports on his forehead two small feathered horns which have won for him in the district the name of Machoto banarudo, the Horned Owl. His song, which is rich enough to fill by itself the still night air, is of a nerve-shattering monotony. With imperturbable and measured regularity, for hours on end, "kew, kew," the bird spits out its cantata to the moon.

One of them has arrived at this moment, driven from the plane-trees in the square by the din of the rejoicings, to demand my hospitality. I can hear him in the top of a cypress near by. From up there, dominating the lyrical assembly, at regular intervals he cuts into the vague orchestration of the Grasshoppers and the Toads.

His soft note is contrasted, intermittently, with a sort of Cat's mew, coming from another spot. This is the call of the Common Owl, the meditative bird of Minerva. After hiding all day in the seclusion of a hollow olive-tree, he started on his wanderings when the shades of evening began to fall. Swinging along with a sinuous flight, he came from somewhere in the neighbourhood to the pines in my enclosure, whence he mingles his harsh mewing, slightly softened by distance, with the general concert.

The Green Grasshopper's clicking is too faint to be clearly perceived amidst these clamourers; all that reaches me is the least ripple, just noticeable when there is a moment's silence. He possesses as his apparatus of sound only a modest drum and scraper, whereas they, more highly privileged, have their bellows, the lungs, which send forth a column of vibrating air. There is no comparison possible. Let us return to the insects.

One of these, though inferior in size and no less sparingly equipped, greatly surpasses the Grasshopper in nocturnal rhapsodies. I speak of the pale and slender Italian Cricket (Oecanthus pellucens, Scop.), who is so puny that you dare not take him up for fear of crushing him. He makes music everywhere among the rosemary-bushes, while the Glow-worms light up their blue lamps to complete the revels. The delicate instrumentalist consists chiefly of a pair of large wings, thin and gleaming as strips of mica. Thanks to these dry sails, he fiddles away with an intensity capable of drowning the Toads' fugue. His performance suggests, but with more brilliancy, more tremolo in the execution, the song of the Common Black Cricket. Indeed the mistake would certainly be made by any one who did not know that, by the time the very hot weather comes, the true Cricket, the chorister of spring, has disappeared. His pleasant violin has been succeeded by another more pleasant still and worthy of special study. We shall return to him at an opportune moment.

These then, limiting ourselves to select specimens, are the principal participants in this musical evening: the Scops-owl, with his languorous solos; the Toad, that tinkler of sonatas; the Italian Cricket, who scrapes the first string of a violin; and the Green Grasshopper, who seems to beat a tiny steel triangle.

We are celebrating to-day, with greater uproar than conviction, the new era, dating politically from the fall of the Bastille; they, with glorious indifference to human things, are celebrating the festival of the sun, singing the happiness of existence, sounding the loud hosanna of the July heats.

What care they for man and his fickle rejoicings! For whom or for what will our squibs be spluttering a few years hence? Far-seeing indeed would he be who could answer the question. Fashions change and bring us the unexpected. The time-serving rocket spreads its sheaf of sparks for the public enemy of yesterday, who has become the idol of to-day. Tomorrow it will go up for somebody else.

In a century or two, will any one, outside the historians, give a thought to the taking of the Bastille? It is very doubtful. We shall have other joys and also other cares.

Let us look a little farther ahead. A day will come, so everything seems to tell us, when, after making progress upon progress, man will succumb, destroyed by the excess of what he calls civilization. Too eager to play the god, he cannot hope for the animal's placid longevity; he will have disappeared when the little Toad is still saying his litany, in company with the Grasshopper, the Scops-owl and the others. They were singing on this planet before us; they will sing after us, celebrating what can never change, the fiery glory of the sun.

I will dwell no longer on this festival and will become once more the naturalist, anxious to obtain information concerning the private life of the insect. The Green Grasshopper (Locusta viridissima, Lin.) does not appear to be common in my neighbourhood. Last year, intending to make a study of this insect and finding my efforts to hunt it fruitless, I was obliged to have recourse to the good offices of a forest-ranger, who sent me a pair of couples from the Lagarde plateau, that bleak district where the beech-tree begins its escalade of the Ventoux.

Now and then freakish fortune takes it into her head to smile upon the persevering. What was not to be found last year has become almost common this summer. Without leaving my narrow enclosure, I obtain as many Grasshoppers as I could wish. I hear them rustling at night in the green thickets. Let us make the most of the windfall, which perhaps will not occur again.

In the month of June my treasures are installed, in a sufficient number of couples, under a wire cover standing on a bed of sand in an earthen pan. It is indeed a magnificent insect, pale-green all over, with two whitish stripes running down its sides. Its imposing size, its slim proportions and its great gauze wings make it the most elegant of our Locustidae. I am enraptured with my captives. What will they teach me? We shall see. For the moment, we must feed them.

I offer the prisoners a leaf of lettuce. They bite into it, certainly, but very sparingly and with a scornful tooth. It soon becomes plain that I am dealing with half-hearted vegetarians. They want something else: they are beasts of prey, apparently. But what manner of prey? A lucky chance taught me.

At break of day I was pacing up and down outside my door, when something fell from the nearest plane-tree with a shrill grating sound. I ran up and saw a Grasshopper gutting the belly of a struggling Cicada. In vain the victim buzzed and waved his limbs: the other did not let go, dipping her head right into the entrails and rooting them out by small mouthfuls.

I knew what I wanted to know: the attack had taken place up above, early in the morning, while the Cicada was asleep; and the plunging of the poor wretch, dissected alive, had made assailant and assailed fall in a bundle to the ground. Since then I have repeatedly had occasion to witness similar carnage.

I have even seen the Grasshopper—the height of audacity, this—dart in pursuit of a Cicada in mad flight. Even so does the Sparrow-hawk pursue the Swallow in the sky. But the bird of prey here is inferior to the insect. It attacks a weaker than itself. The Grasshopper, on the other hand, assaults a colossus, much larger than herself and stronger; and nevertheless the result of the unequal fight is not in doubt. The Grasshopper rarely fails with the sharp pliers of her powerful jaws to disembowel her capture, which, being unprovided with weapons, confines itself to crying out and kicking.

The main thing is to retain one's hold of the prize, which is not difficult in somnolent darkness. Any Cicada encountered by the fierce Locustid on her nocturnal rounds is bound to die a lamentable death. This explains those sudden agonized notes which grate through the woods at late, unseasonable hours, when the cymbals have long been silent. The murderess in her suit of apple-green has pounced on some sleeping Cicada.

My boarders' menu is settled: I will feed them on Cicadae. They take such a liking to this fare that, in two or three weeks, the floor of the cage is a knacker's yard strewn with heads and empty thoraces, with torn-off wings and disjointed legs. The belly alone disappears almost entirely. This is the tit-bit, not very substantial, but extremely tasty, it would seem. Here, in fact, in the insect's crop, the syrup is accumulated, the sugary sap which the Cicada's gimlet taps from the tender bark. Is it because of this dainty that the prey's abdomen is preferred to any other morsel? It is quite possible.

I do, in fact, with a view to varying the diet, decide to serve up some very sweet fruits, slices of pear, grape-bits, bits of melon. All this meets with delighted appreciation. The Green Grasshopper resembles the English: she dotes on underdone meat seasoned with jelly. This perhaps is why, on catching the Cicada, she first rips up his paunch, which supplies a mixture of flesh and preserves.

To eat Cicadae and sugar is not possible in every part of the country. In the north, where she abounds, the Green Grasshopper would not find the dish which attracts her so strongly here. She must have other resources. To convince myself of this, I give her Anoxiae (A. pilosa, Fab.), the summer equivalent of the spring Cockchafer. The Beetle is accepted without hesitation. Nothing is left of him but the wing-cases, head and legs. The result is the same with the magnificent plump Pine Cockchafer (Melolontha fullo, Lin.), a sumptuous morsel which I find next day eviscerated by my gang of knackers.

These examples teach us enough. They tell us that the Grasshopper is an inveterate consumer of insects, especially of those which are not protected by too hard a cuirass; they are evidence of tastes which are highly carnivorous, but not exclusively so, like those of the Praying Mantis, who refuses everything except game. The butcher of the Cicadae is able to modify an excessively heating diet with vegetable fare. After meat and blood, sugary fruit-pulp; sometimes even, for lack of anything better, a little green stuff.

Nevertheless, cannibalism is prevalent. True, I never witness in my Grasshopper-cages the savagery which is so common in the Praying Mantis, who harpoons her

rivals and devours her lovers; but, if some weakling succumb, the survivors hardly ever fail to profit by his carcass as they would in the case of any ordinary prey.
With no scarcity of provisions as an excuse, they feast upon their defunct companion. For the rest, all the sabre-bearing clan display, in varying degrees, a
propensity for filling their bellies with their maimed comrades.

In other respects, the Grasshoppers live together very peacefully in my cages. No
serious strife ever takes place among them, nothing beyond a little rivalry in the
matter of food. I hand in a piece of pear. A Grasshopper alights on it at once.
Jealously she kicks away any one trying to bite at the delicious morsel. Selfishness
reigns everywhere. When she has eaten her fill, she makes way for another, who
in her turn becomes intolerant. One after the other, all the inmates of the
menagerie come and refresh themselves. After cramming their crops, they scratch
the soles of their feet a little with their mandibles, polish up their forehead and
eyes with a leg moistened with spittle and then, hanging to the trellis-work or
lying on the sand in a posture of contemplation, blissfully they digest and slum-
ber most of the day, especially during the hottest part of it.

It is in the evening, after sunset, that the troop becomes lively. By nine o'clock the
animation is at its height. With sudden rushes they clamber to the top of the
dome, to descend as hurriedly and climb up once more. They come and go
tumultuously, run and hop around the circular track and, without stopping, nibble at the good things on the way.

The males are stridulating by themselves, here and there, teasing the passing fair
with their antennae. The future mothers stroll about gravely, with their sabre half-
raised. The agitation and feverish excitement means that the great business of
pairing is at hand. The fact will escape no practised eye.

It is also what I particularly wish to observe. My wish is satisfied, but not fully,
for the late hours at which events take place did not allow me to witness the final
act of the wedding. It is late at night or early in the morning that things happen.

The little that I see is confined to interminable preludes. Standing face to face,
with foreheads almost touching, the lovers feel and sound each other for a long
time with their limp antennae. They suggest two fencers crossing and recrossing
harmless foils. From time to time, the male stridulates a little, gives a few short
strokes of the bow and then falls silent, feeling perhaps too much overcome to
continue. Eleven o'clock strikes; and the declaration is not yet over. Very regret-
fully, but conquered by sleepiness, I quit the couple.

Next morning, early, the female carries, hanging at the bottom of her ovipositor, a queer bladder-like arrangement, an opaline capsule, the size of a large pea and roughly subdivided into a small number of egg-shaped vesicles. When the insect walks, the thing scrapes along the ground and becomes dirty with sticky grains of sand. The Grasshopper then makes a banquet off this fertilizing capsule, drains it slowly of its contents, and devours it bit by bit; for a long time she chews and rechews the gummy morsel and ends by swallowing it all down. In less than half a day, the milky burden has disappeared, consumed with zest down to the last atom.

This inconceivable banquet must be imported, one would think, from another planet, so far removed is it from earthly habits. What a singular race are the Locustidae, one of the oldest in the animal kingdom on dry land and, like the Scolopendra and the Cephalopod, acting as a belated representative of the manners of antiquity!

CHAPTER 3.

THE EMPUSA.

The sea, life's first foster-mother, still preserves in her depths many of those singular and incongruous shapes which were the earliest attempts of the animal kingdom; the land, less fruitful, but with more capacity for progress, has almost wholly lost the strange forms of other days. The few that remain belong especially to the series of primitive insects, insects exceedingly limited in their industrial powers and subject to very summary metamorphoses, if to any at all. In my district, in the front rank of those entomological anomalies which remind us of the denizens of the old coal-forests, stand the Mantidae, including the Praying Mantis, so curious in habits and structure. Here also is the Empusa (E. pauperata, Latr.), the subject of this chapter.

Her larva is certainly the strangest creature among the terrestrial fauna of Provence: a slim, swaying thing of so fantastic an appearance that uninitiated fingers dare not lay hold of it. The children of my neighbourhood, impressed by its startling shape, call it "the Devilkin." In their imaginations, the queer little creature savours of witchcraft. One comes across it, though always sparsely, in spring, up to May; in autumn; and sometimes in winter, if the sun be strong. The tough grasses of the waste-lands, the stunted bushes which catch the sun and are sheltered from the wind by a few heaps of stones are the chilly Empusa's favourite abode.

Let us give a rapid sketch of her. The abdomen, which always curls up so as to join the back, spreads paddle wise and twists into a crook. Pointed scales, a sort of foliaceous expansions arranged in three rows, cover the lower surface, which becomes the upper surface because of the crook aforesaid. The scaly crook is propped on four long, thin stilts, on four legs armed with knee-pieces, that is to say, carrying at the end of the thigh, where it joins the shin, a curved, projecting blade not unlike that of a cleaver.

Above this base, this four-legged stool, rises, at a sudden angle, the stiff corselet, disproportionately long and almost perpendicular. The end of this bust, round and slender as a straw, carries the hunting-trap, the grappling limbs, copied from those of the Mantis. They consist of a terminal harpoon, sharper than a needle, and a cruel vice, with the jaws toothed like a saw. The jaw formed by the arm proper is hollowed into a groove and carries on either side five long spikes, with smaller indentations in between. The jaw formed by the forearm is similarly furrowed, but its double saw, which fits into the groove of the upper arm when at rest, is formed of finer, closer and more regular teeth. The magnifying-glass reveals a score of equal points in each row. The machine only lacks size to be a fearful implement of torture.

The head is in keeping with this arsenal. What a queer-shaped head it is! A pointed face, with walrus moustaches furnished by the palpi; large goggle eyes; between them, a dirk, a halberd blade; and, on the forehead a mad, unheard of thing: a sort of tall mitre, an extravagant head-dress that juts forward, spreading right and left into peaked wings and cleft along the top. What does the Devilkin want with that monstrous pointed cap, than which no wise man of the East, no astrologer of old ever wore a more splendiferous? This we shall learn when we see her out hunting.

The dress is commonplace; grey tints predominate. Towards the end of the larval period, after a few moultings, it begins to give a glimpse of the adult's richer livery and becomes striped, still very faintly, with pale-green, white and pink. Already the two sexes are distinguished by their antennae. Those of the future mothers are thread-like; those of the future males are distended into a spindle at the lower half, forming a case or sheath whence graceful plumes will spring at a later date.

Behold the creature, worthy of a Callot's fantastic pencil. (Jacques Callot (1592-1635), the French engraver and painter, famed for the grotesque nature of his subjects.—Translator's Note.) If you come across it in the bramble-bushes, it sways upon its four stilts, it wags its head, it looks at you with a knowing air, it twists its mitre round and peers over its shoulder. You seem to read mischief in its pointed face. You try to take hold of it. The imposing attitude ceases forthwith, the raised corselet is lowered and the creature makes off with mighty strides, helping itself along with its fighting-limbs, which clutch the twigs. The flight need not last long, if you have a practised eye. The Empusa is captured, put into a screw of paper, which will save her frail limbs from sprains, and lastly penned in a wire-gauze cage. In this way, in October, I obtain a flock sufficient for my purpose.

How to feed them? My Devilkins are very little; they are a month or two old at most. I give them Locusts suited to their size, the smallest that I can find. They refuse them. Nay more, they are frightened of them. Should a thoughtless Locust meekly approach one of the Empusae, suspended by her four hind-legs to the trellised dome, the intruder meets with a bad reception. The pointed mitre is lowered; and an angry thrust sends him rolling. We have it: the wizard's cap is a defensive weapon, a protective crest. The Ram charges with his forehead, the Empusa butts with her mitre.

But this does not mean dinner. I serve up the House-fly, alive. She is accepted, without hesitation. The moment that the Fly comes within reach, the watchful Devilkin turns her head, bends the stalk of her corselet slantwise and, flinging out her fore-limb, harpoons the Fly and grips her between her two saws. No Cat pouncing upon a Mouse could be quicker.

The game, however small, is enough for a meal. It is enough for the whole day, often for several days. This is my first surprise: the extreme abstemiousness of these fiercely-armed insects. I was prepared for ogres: I find ascetics satisfied with a meagre collation at rare intervals. A Fly fills their belly for twenty-four hours at least.

Thus passes the late autumn: the Empusae, more and more temperate from day to day, hang motionless from the wire gauze. Their natural abstinence is my best ally, for Flies grow scarce; and a time comes when I should be hard put to it to keep the menageries supplied with provisions.

During the three winter months, nothing stirs. From time to time, on fine days, I expose the cage to the sun's rays, in the window. Under the influence of this heat-bath, the captives stretch their legs a little, sway from side to side, make up their minds to move about, but without displaying any awakening appetite. The rare Midges that fall to my assiduous efforts do not appear to tempt them. It is a rule for them to spend the cold season in a state of complete abstinence.

My cages tell me what must happen outside, during the winter. Ensconced in the crannies of the rockwork, in the sunniest places, the young Empusae wait, in a state of torpor, for the return of the hot weather. Notwithstanding the shelter of a heap of stones, there must be painful moments when the frost is prolonged and the snow penetrates little by little into the best-protected crevices. No matter: hardier than they look, the refugees escape the dangers of the winter season. Sometimes, when the sun is strong, they venture out of their hiding-place and come to see if spring be nigh.

Spring comes. We are in March. My prisoners bestir themselves, change their skin. They need victuals. My catering difficulties recommence. The House-fly, so easy to catch, is lacking in these days. I fall back upon earlier Diptera: Eristales, or Drone-flies. The Empusa refuses them. They are too big for her and can offer too strenuous a resistance. She wards off their approach with blows of her mitre.

A few tender morsels, in the shape of very young Grasshoppers, are readily accepted. Unfortunately, such windfalls do not often find their way into my sweeping-net. Abstinence becomes obligatory until the arrival of the first Butterflies. Henceforth, Pieris brassicae, the White Cabbage Butterfly, will contribute the greater portion of the victuals.

Let loose in the wire cage, the Pieris is regarded as excellent game. The Empusa lies in wait for her, seizes her, but releases her at once, lacking the strength to over-power her. The Butterfly's great wings, beating the air, give her shock after shock and compel her to let go. I come to the weakling's assistance and cut the wings of her prey with my scissors. The maimed ones, still full of life, clamber up the trel-lis-work and are forthwith grabbed by the Empusae, who, in no way frightened by their protests, crunch them up. The dish is to their taste and, moreover, plen-tiful, so much so that there are always some despised remnants.

The head only and the upper portion of the breast are devoured: the rest—the plump abdomen, the best part of the thorax, the legs and lastly, of course, the wing-stumps—is flung aside untouched. Does this mean that the tenderest and most succulent morsels are chosen? No, for the belly is certainly more juicy; and the Empusa refuses it, though she eats up her House-fly to the last particle. It is a strategy of war. I am again in the presence of a neck-specialist as expert as the Mantis herself in the art of swiftly slaying a victim that struggles and, in strug-gling, spoils the meal.

Once warned, I soon perceive that the game, be it Fly, Locust, Grasshopper, or Butterfly, is always struck in the neck, from behind. The first bite is aimed at the point containing the cervical ganglia and produces sudden death or immobility. Complete inertia will leave the consumer in peace, the essential condition of every satisfactory repast.

The Devilkin, therefore, frail though she be, possesses the secret of immediately destroying the resistance of her prey. She bites at the back of the neck first, in order to give the finishing stroke. She goes on nibbling around the original attacking-point. In this way the Butterfly's head and the upper part of the breast are disposed of. But, by that time, the huntress is surfeited: she wants so little! The rest lies on the ground, disdained, not for lack of flavour, but because there is too much of it. A Cabbage Butterfly far exceeds the capacity of the Empusa's stomach. The Ants will benefit by what is left.

There is one other matter to be mentioned, before observing the metamorphosis. The position adopted by the young Empusae in the wire-gauze cage is invariably the same from start to finish. Gripping the trellis-work by the claws of its four hind-legs, the insect occupies the top of the dome and hangs motionless, back downwards, with the whole of its body supported by the four suspension-points. If it wishes to move, the front harpoons open, stretch out, grasp a mesh and draw it to them. When the short walk is over, the lethal arms are brought back against the chest. One may say that it is nearly always the four hind-shanks which alone support the suspended insect.

And this reversed position, which seems to us so trying, lasts for no short while: it is prolonged, in my cages, for ten months without a break. The Fly on the ceiling, it is true, occupies the same attitude; but she has her moments of rest: she flies, she walks in a normal posture, she spreads herself flat in the sun. Besides, her acrobatic feats do not cover a long period. The Empusa, on the other hand, maintains her curious equilibrium for ten months on end, without a break. Hanging from the trellis-work, back downwards, she hunts, eats, digests, dozes, casts her skin, undergoes her transformation, mates, lays her eggs and dies. She clambered up there when she was still quite young; she falls down, full of days, a corpse.

Things do not happen exactly like this under natural conditions. The insect stands on the bushes back upwards; it keeps its balance in the regular attitude and turns over only in circumstances that occur at long intervals. The protracted suspension of my captives is all the more remarkable inasmuch as it is not at all an innate habit of their race.

It reminds one of the Bats, who hang, head downwards, by their hind-legs from the roof of their caves. A special formation of the toes enables birds to sleep on one leg, which automatically and without fatigue clutches the swaying bough. The Empusa shows me nothing akin to their contrivance. The extremity of her walking-legs has the ordinary structure: a double claw at the tip, a double steel-yard-hook; and that is all.

I could wish that anatomy would show me the working of the muscles and nerves in those tarsi, in those legs more slender than threads, the action of the tendons that control the claws and keep them gripped for ten months, unwearied in waking and sleeping. If some dexterous scalpel should ever investigate this problem, I can recommend another, even more singular than that of the Empusa, the Bat and the bird. I refer to the attitude of certain Wasps and Bees during the night's rest.

An Ammophila with red fore-legs (A. holosericea) is plentiful in my enclosure towards the end of August and selects a certain lavender-border for her dormitory. At dusk, especially after a stifling day, when a storm is brewing, I am sure to find the strange sleeper settled there. Never was more eccentric attitude adopted for a night's rest! The mandibles bite right into the lavender-stem. Its square shape supplies a firmer hold than a round stalk would do. With this one and only prop, the animal's body juts out stiffly, at full length, with legs folded. It forms a right angle with the supporting axis, so much so that the whole weight of the insect, which has turned itself into the arm of a lever rests upon the mandibles.

The Ammophila sleeps extended in space by virtue of her mighty jaws. It takes an animal to think of a thing like that, which upsets all our preconceived ideas of repose. Should the threatening storm burst, should the stalk sway in the wind, the sleeper is not troubled by her swinging hammock; at most, she presses her fore-legs for a moment against the tossed mast. As soon as equilibrium is restored, the favourite posture, that of the horizontal lever, is resumed. perhaps the mandibles, like the bird's toes, possess the faculty of gripping tighter in proportion to the rocking of the wind.

The Ammophila is not the only one to sleep in this singular position, which is copied by many others—Anthidia (Cotton-bees.—Translator's Note.), Odyneri (A genus of Mason-wasps.—Translator's Note.), Eucerae (A species of Burrowing-bees.—Translator's Note.)—and mainly by the males. All grip a stalk with their mandibles and sleep with their bodies outstretched and their legs folded back. Some, the stouter species, allow themselves to rest the tip of their arched abdomen against the pole.

This visit to the dormitory of certain Wasps and Bees does not explain the problem of the Empusa; it sets up another one, no less difficult. It shows us how deficient we are in insight, when it comes to differentiating between fatigue and rest in the cogs of the animal machine. The Ammophila, with the static paradox afforded by her mandibles; the Empusa, with her claws unwearied by ten months' hanging, leave the physiologist perplexed and make him wonder what really constitutes rest. In absolute fact, there is no rest, apart from that which puts an end to life. The struggle never ceases; some muscle is always toiling, some nerve strain-ing. Sleep, which resembles a return to the peace of non-existence, is, like waking, an effort, here of the leg, of the curled tail; there of the claw, of the jaws.

The transformation is effected about the middle of May, and the adult Empusa makes her appearance. She is even more remarkable in figure and attire than the Praying Mantis. Of her youthful eccentricities, she retains the pointed mitre, the saw-like arm-guards, the long bust, the knee-pieces, the three rows of scales on the lower surface of the belly; but the abdomen is now no longer twisted into a crook and the animal is comelier to look upon. Large pale-green wings, pink at the shoulder and swift in flight in both sexes, cover the belly, which is striped white and green underneath. The male, the dandy sex, adorns himself with plumed antennae, like those of certain Moths, the Bombyx tribe. In respect of size, he is almost the equal of his mate.

Save for a few slight structural details, the Empusa is the Praying Mantis. The peasant confuses them. When, in spring, he meets the mitred insect, he thinks he sees the common PrŠgo-Dieu, who is a daughter of the autumn. Similar forms would seem to indicate similarity of habits. In fact, led away by the extraordinary armour, we should be tempted to attribute to the Empusa a mode of life even more atrocious than that of the Mantis. I myself thought so at first; and any one, relying upon false analogies, would think the same. It is a fresh error: for all her warlike aspect, the Empusa is a peaceful creature that hardly repays the trouble of rearing.

Installed under the gauze bell, whether in assemblies of half a dozen or in sepa-rate couples, she at no time loses her placidity. Like the larva, she is very abstemious and contents herself with a Fly or two as her daily ration.

Big eaters are naturally quarrelsome. The Mantis, bloated with Locusts, soon becomes irritated and shows fight. The Empusa, with her frugal meals, does not indulge in hostile demonstrations. There is no strife among neighbours nor any of those sudden unfurlings of the wings so dear to the Mantis when she assumes the spectral attitude and puffs like a startled Adder; never the least inclination for those cannibal banquets whereat the sister who has been worsted in the fight is devoured. Such atrocities or here unknown.

Unknown also are tragic nuptials. The male is enterprising and assiduous and is subjected to a long trial before succeeding. For days and days he worries his mate, who ends by yielding. Due decorum is preserved after the wedding. The feathered groom retires, respected by his bride, and does his little bit of hunting, without danger of being apprehended and gobbled up.

The two sexes live together in peace and mutual indifference until the middle of July. Then the male, grown old and decrepit, takes counsel with himself, hunts no more, becomes shaky in his walk, creeps down from the lofty heights of the trellised dome and at last collapses on the ground. His end comes by a natural death. And remember that the other, the male of the Praying Mantis, ends in the stomach of his gluttonous spouse.

The laying follows close upon the disappearance of the males.

One word more on comparative manners. The Mantis goes in for battle and cannibalism; the Empusa is peaceable and respects her kind. To what cause are these profound moral differences due, when the organic structure is the same? Perhaps to the difference of diet. Frugality, in fact, softens character, in animals as in men; gross feeding brutalizes it. The gormandizer gorged with meat and strong drink, a fruitful source of savage outbursts, could not possess the gentleness of the ascetic who dips his bread into a cup of milk. The Mantis is that gormandizer, the Empusa that ascetic.

Granted. But whence does the one derive her voracious appetite, the other her temperate ways, when it would seem as though their almost identical structure ought to produce an identity of needs? These insects tell us, in their fashion, what many have already told us: that propensities and aptitudes do not depend exclusively upon anatomy; high above the physical laws that govern matter rise other laws that govern instincts.

CHAPTER 4.

THE CAPRICORN.

My youthful meditations owe some happy moments to Condillac's famous statue which, when endowed with the sense of smell, inhales the scent of a rose and out of that single impression creates a whole world of ideas. (Etienne Bonnot de Condillac, Abb, de Mureaux (1715-80), the leading exponent of sensational philosophy. His most important work is the "Trait, des sensations," in which he imagines a statue, organized like a man, and endows it with the senses one by one, beginning with that of smell. He argues by a process of imaginative reconstruction that all human faculties and all human knowledge are merely transformed sensation, to the exclusion of any other principle, that, in short, everything has its source in sensation: man is nothing but what he has acquired.—Translator's Note.) My twenty-year-old mind, full of faith in syllogisms, loved to follow the deductive jugglery of the abb,-philosopher: I saw, or seemed to see, the statue take life in that action of the nostrils, acquiring attention, memory, judgment and all the psychological paraphernalia, even as still waters are aroused and rippled by the impact of a grain of sand. I recovered from my illusion under the instruction of my abler master, the animal. The Capricorn shall teach us that the problem is more obscure than the abb, led me to believe.

When wedge and mallet are at work, preparing my provision of firewood under the grey sky that heralds winter, a favourite relaxation creates a welcome break in my daily output of prose. By my express orders, the woodman has selected the oldest and most ravaged trunks in his stack. My tastes bring a smile to his lips; he wonders by what whimsy I prefer wood that is worm-eaten—chirouna, as he calls it—to sound wood which burns so much better. I have my views on the subject; and the worthy man submits to them.

And now to us two, O my fine oak-trunk seamed with scars, gashed with wounds whence trickle the brown drops smelling of the tan-yard. The mallet drives home, the wedges bite, the wood splits. What do your flanks contain? Real treasures for my studies. In the dry and hollow parts, groups of various insects, capable of living through the bad season of the year, have taken up their winter quarters: in the low-roofed galleries, galleries which some Buprestis-beetle has built, Osmia-bees, working their paste of masticated leaves, have piled their cells, one above the other; in the deserted chambers and vestibules, Megachiles (Leaf-cutting Bees.— Translator's Note.) have arranged their leafy jars; in the live wood, filled with juicy saps, the larvae of the Capricorn (Cerambyx miles), the chief author of the oak's undoing, have set up their home.

Strange creatures, of a verity, are these grubs, for an insect of superior organization: bits of intestines crawling about! At this time of year, the middle of autumn, I meet them of two different ages. The older are almost as thick as one's finger; the others hardly attain the diameter of a pencil. I find, in addition, pupae more or less fully coloured, perfect insects, with a distended abdomen, ready to leave the trunk when the hot weather comes again. Life inside the wood, therefore, lasts three years. How is this long period of solitude and captivity spent? In wandering lazily through the thickness of the oak, in making roads whose rubbish serves as food. The horse in Job swallows the ground in a figure of speech; the Capricorn's grub literally eats its way. ("Chafing and raging, he swalloweth the ground, neither doth he make account when the noise of the trumpet soundeth."—Job 39, 23 (Douai version).—Translator's Note.) With its carpenter's gouge, a strong black mandible, short, devoid of notches, scooped into a sharp-edged spoon, it digs the opening of its tunnel. The piece cut out is a mouthful which, as it enters the stomach, yields its scanty juices and accumulates behind the worker in heaps of wormed wood. The refuse leaves room in front by passing through the worker. A labour at once of nutrition and of road-making, the path is devoured while constructed; it is blocked behind as it makes way ahead. That, however, is how all the borers who look to wood for victuals and lodging set about their business.

For the harsh work of its two gouges, or curved chisels, the larva of the Capricorn concentrates its muscular strength in the front of its body, which swells into a pestle-head. The Buprestis-grubs, those other industrious carpenters, adopt a similar form; they even exaggerate their pestle. The part that toils and carves hard wood requires a robust structure; the rest of the body, which has but to follow after, continues slim. The essential thing is that the implement of the jaws should possess a solid support and a powerful motor. The Cerambyx-larva strengthens its chisels with a stout, black, horny armour that surrounds the mouth; yet, apart from its skull and its equipment of tools, the grub has a skin as fine as satin and white as ivory. This dead white comes from a copious layer of grease which the animal's spare diet would not lead us to suspect. True, it has nothing to do, at every hour of the day and night, but gnaw. The quantity of wood that passes into its stomach makes up for the dearth of nourishing elements.

The legs, consisting of three pieces, the first globular, the last sharp-pointed, are mere rudiments, vestiges. They are hardly a millimetre long. (.039 inch.— Translator's Note.) For this reason they are of no use whatever for walking; they do not even bear upon the supporting surface, being kept off it by the obesity of the chest. The organs of locomotion are something altogether different. The grub of the Capricorn moves at the same time on its back and belly; instead of the useless legs of the thorax, it has a walking-apparatus almost resembling feet, which appear, contrary to every rule, on the dorsal surface.

The first seven segments of the abdomen have, both above and below, a four-sided facet, bristling with rough protuberances. This the grub can either expand or contract, making it stick out or lie flat at will. The upper facets consist of two excrescences separated by the mid-dorsal line; the lower ones have not this divided appearance. These are the organs of locomotion, the ambulacra. When the larva wishes to move forwards, it expands its hinder ambulacra, those on the back as well as those on the belly, and contracts its front ones. Fixed to the side of the narrow gallery by their ridges, the hind-pads give the grub a purchase. The flattening of the fore-pads, by decreasing the diameter, allows it to slip forward and to take half a step. To complete the step the hind-quarters have to be brought up the same distance. With this object, the front pads fill out and provide support, while those behind shrink and leave free scope for their segments to contract.

With the double support of its back and belly, with alternate puffings and shrinkings, the animal easily advances or retreats along its gallery, a sort of mould which the contents fill without a gap. But if the locomotory pads grip only on one side progress becomes impossible. When placed on the smooth wood of my table, the animal wriggles slowly; it lengthens and shortens without advancing by a hair's-breadth. Laid on the surface of a piece of split oak, a rough, uneven surface, due to the gash made by the wedge, it twists and writhes, moves the front part of its body very slowly from left to right and right to left, lifts it a little, lowers it and begins again. These are the most extensive movements made. The vestigial legs remain inert and absolutely useless. Then why are they there? It were better to lose them altogether, if it be true that crawling inside the oak has deprived the animal of the good legs with which it started. The influence of environment, so well-inspired in endowing the grub with ambulatory pads, becomes a mockery when it leaves it these ridiculous stumps. Can the structure, perchance, be obeying other rules than those of environment?

Though the useless legs, the germs of the future limbs, persist, there is no sign in the grub of the eyes wherewith the Cerambyx will be richly gifted. The larva has not the least trace of organs of vision. What would it do with sight in the murky thickness of a tree-trunk? Hearing is likewise absent. In the never-troubled silence of the oak's inmost heart, the sense of hearing would be a non-sense. Where sounds are lacking, of what use is the faculty of discerning them? Should there be any doubts, I will reply to them with the following experiment. Split lengthwise, the grub's abode leaves a half-tunnel wherein I can watch the occupant's doings. When left alone, it now gnaws the front of its gallery, now rests, fixed by its ambulacra to the two sides of the channel. I avail myself of these moments of quiet to inquire into its power of perceiving sounds. The banging of hard bodies, the ring of metallic objects, the grating of a file upon a saw are tried in vain. The animal remains impassive. Not a wince, not a movement of the skin; no sign of awakened attention. I succeed no better when I scratch the wood close by with a hard point, to imitate the sound of some neighbouring larva gnawing the intervening thickness. The indifference to my noisy tricks could be no greater in a lifeless object. The animal is deaf.

Can it smell? Everything tells us no. Scent is of assistance in the search for food. But the Capricorn grub need not go in quest of eatables: it feeds on its home, it lives on the wood that gives it shelter. Let us make an attempt or two, however. I scoop in a log of fresh cypress-wood a groove of the same diameter as that of the natural galleries and I place the worm inside it. Cypress-wood is strongly scented; it possesses in a high degree that resinous aroma which characterizes most of the pine family. Well, when laid in the odoriferous channel, the larva goes to the end, as far as it can go, and makes no further movement. Does not this placid quiescence point to the absence of a sense of smell? The resinous flavour, so strange to the grub which has always lived in oak, ought to vex it, to trouble it; and the disagreeable impression ought to be revealed by a certain commotion, by certain attempts to get away. Well, nothing of the kind happens: once the larva has found the right position in the groove, it does not stir. I do more: I set before it, at a very short distance, in its normal canal, a piece of camphor. Again, no effect. Camphor is followed by naphthaline. Still nothing. After these fruitless endeavours, I do not think that I am going too far when I deny the creature a sense of smell.

Taste is there, no doubt. But such taste! The food is without variety: oak, for three years at a stretch, and nothing else. What can the grub's palate appreciate in this

monotonous fare? The tannic relish of a fresh piece, oozing with sap, the uninteresting flavour of an over-dry piece, robbed of its natural condiment: these
probably represent the whole gustative scale.

There remains touch, the far-spreading, passive sense common to all live flesh that
quivers under the goad of pain. The sensitive schedule of the Cerambyx-grub,
therefore, is limited to taste and touch, both exceedingly obtuse. This almost
brings us to Condillac's statue. The imaginary being of the philosopher had one
sense only, that of smell, equal in delicacy to our own; the real being, the ravager
of the oak, has two, inferior, even when put together, to the former, which so
plainly perceived the scent of a rose and distinguished it so clearly from any other.
The real case will bear comparison with the fictitious.

What can be the psychology of a creature possessing such a powerful digestive
organism combined with such a feeble set of senses? A vain wish has often come
to me in my dreams; it is to be able to think, for a few minutes, with the crude
brain of my Dog, to see the world with the faceted eyes of a Gnat. How things
would change in appearance! They would change much more if interpreted by the
intellect of the grub. What have the lessons of touch and taste contributed to that
rudimentary receptacle of impressions? Very little; almost nothing. The animal
knows that the best bits possess an astringent flavour; that the sides of a passage
not carefully planed are painful to the skin. This is the utmost limit of its acquired
wisdom. In comparison, the statue with the sensitive nostrils was a marvel of
knowledge, a paragon too generously endowed by its inventor. It remembered,
compared, judged, reasoned: does the drowsily digesting paunch remember? Does
it compare? Does it reason? I defined the Capricorn-grub as a bit of an intestine
that crawls about. The undeniable accuracy of this definition provides me with
my answer: the grub has the aggregate of sense-impressions that a bit of an intestine may hope to have.

And this nothing-at-all is capable of marvellous acts of foresight; this belly, which
knows hardly aught of the present, sees very clearly into the future. Let us take an
illustration on this curious subject. For three years on end the larva wanders about
in the thick of the trunk; it goes up, goes down, turns to this side and that; it
leaves one vein for another of better flavour, but without moving too far from the
inner depths, where the temperature is milder and greater safety reigns. A day is
at hand, a dangerous day for the recluse obliged to quit its excellent retreat and
face the perils of the surface. Eating is not everything: we have to get out of this.
The larva, so well-equipped with tools and muscular strength, finds no difficulty
in going where it pleases, by boring through the wood; but does the coming
Capricorn, whose short spell of life must be spent in the open air, possess the same

advantages? Hatched inside the trunk, will the long-horned insect be able to clear itself a way of escape?

That is the difficulty which the worm solves by inspiration. Less versed in things of the future, despite my gleams of reason, I resort to experiment with a view to fathoming the question. I begin by ascertaining that the Capricorn, when he wishes to leave the trunk, is absolutely unable to make use of the tunnel wrought by the larva. It is a very long and very irregular maze, blocked with great heaps of wormed wood. Its diameter decreases progressively from the final blind alley to the starting-point. The larva entered the timber as slim as a tiny bit of straw; it is to-day as thick as my finger. In its three years' wanderings it always dug its gallery according to the mould of its body. Evidently, the road by which the larva entered and moved about cannot be the Capricorn's exit-way: his immoderate antennae, his long legs, his inflexible armour-plates would encounter an insuperable obstacle in the narrow, winding corridor, which would have to be cleared of its wormed wood and, moreover, greatly enlarged. It would be less fatiguing to attack the untouched timber and dig straight ahead. Is the insect capable of doing so? We shall see.

I make some chambers of suitable size in oak logs chopped in two; and each of my artificial cells receives a newly transformed Cerambyx, such as my provisions of firewood supply, when split by the wedge, in October. The two pieces are then joined and kept together with a few bands of wire. June comes. I hear a scraping inside my billets. Will the Capricorns come out, or not? The delivery does not seem difficult to me: there is hardly three-quarters of an inch to pierce. Not one emerges. When all is silence, I open my apparatus. The captives, from first to last, are dead. A vestige of sawdust, less than a pinch of snuff, represents all their work.

I expected more from those sturdy tools, their mandibles. But, as I have said elsewhere, the tool does not make the workman. In spite of their boring-implements, the hermits die in my cases for lack of skill. I subject others to less arduous tests. I enclose them in spacious reed-stumps, equal in diameter to the natal cell. The obstacle to be pierced is the natural diaphragm, a yielding partition two or three millimetres thick. (.078 to .117 inch.—Translator's Note.) Some free themselves; others cannot. The less vibrant ones succumb, stopped by the frail barrier. What would it be if they had to pass through a thickness of oak?

We are now persuaded: despite his stalwart appearance, the Capricorn is powerless to leave the tree-trunk by his unaided efforts. It therefore falls to the worm, to the wisdom of that bit of an intestine, to prepare the way for him. We see renewed, in another form, the feats of prowess of the Anthrax, whose pupa, armed

with trepans, bores through rock on the feeble Fly's behalf. Urged by a presenti-
ment that to us remains an unfathomable mystery, the Cerambyx-grub leaves the
inside of the oak, its peaceful retreat, its unassailable stronghold, to wriggle
towards the outside, where lives the foe, the Woodpecker, who may gobble up the
succulent little sausage. At the risk of its life, it stubbornly digs and gnaws to the
very bark, of which it leaves no more intact than the thinnest film, a slender
screen. Sometimes, even, the rash one opens the window wide.

This is the Capricorn's exit-hole. The insect will have but to file the screen a little with its mandibles, to bump against it with its forehead, in order to bring it
down; it will even have nothing to do when the window is free, as often happens.
The unskilled carpenter, burdened with his extravagant head-dress, will emerge
from the darkness through this opening when the summer heats arrive.

After the cares of the future come the cares of the present. The larva, which has
just opened the aperture of escape, retreats some distance down its gallery and, in
the side of the exit-way, digs itself a transformation-chamber more sumptuously
furnished and barricaded than any that I have ever seen. It is a roomy niche,
shaped like a flattened ellipsoid, the length of which reaches eighty to a hundred
millimetres. (3 to 4 inches.—Translator's Note.) The two axes of the cross-section
vary: the horizontal measures twenty-five to thirty millimetres (.975 to 1.17
inch.—Translator's Note.); the vertical measures only fifteen. (.585 inch.—
Translator's Note.) This greater dimension of the cell, where the thickness of the
perfect insect is concerned, leaves a certain scope for the action of its legs when
the time comes for forcing the barricade, which is more than a close-fitting
mummy-case would do.

The barricade in question, a door which the larva builds to exclude the dangers
from without, is two-and even three-fold. Outside, it is a stack of woody refuse,
of particles of chopped timber; inside, a mineral hatch, a concave cover, all in one
piece, of a chalky white. Pretty often, but not always, there is added to these two
layers an inner casing of shavings. Behind this compound door, the larva makes
its arrangements for the metamorphosis. The sides of the chamber are rasped,
thus providing a sort of down formed of ravelled woody fibres, broken into
minute shreds. The velvety matter, as and when obtained, is applied to the wall
in a continuous felt at least a millimetre thick. (.039 inch.—Translator's Note.)
The chamber is thus padded throughout with a fine swan's-down, a delicate precaution taken by the rough worm on behalf of the tender pupa.

Let us hark back to the most curious part of the furnishing, the mineral hatch or
inner door of the entrance. It is an elliptical skull-cap, white and hard as chalk,

smooth within and knotted without, resembling more or less closely an acorn-cup. The knots show that the matter is supplied in small, pasty mouthfuls, solid-ifying outside in slight projections which the insect does not remove, being unable to get at them, and polished on the inside surface, which is within the worm's reach. What can be the nature of that singular lid whereof the Cerambyx furnishes me with the first specimen? It is as hard and brittle as a flake of lime-stone. It can be dissolved cold in nitric acid, discharging little gaseous bubbles. The process of solution is a slow one, requiring several hours for a tiny fragment. Everything is dissolved, except a few yellowish flocks, which appear to be of an organic nature. As a matter of fact, a piece of the hatch, when subjected to heat, blackens, proving the presence of an organic glue cementing the mineral matter. The solution becomes muddy if oxalate of ammonia be added; it then deposits a copious white precipitate. These signs indicate calcium carbonate. I look for urate of ammonia, that constantly recurring product of the various stages of the meta-morphoses. It is not there: I find not the least trace of murexide. The lid, there-fore, is composed solely of carbonate of lime and of an organic cement, no doubt of an albuminous character, which gives consistency to the chalky paste.

Had circumstances served me better, I should have tried to discover in which of the worm's organs the stony deposit dwells. I am however, convinced: it is the stomach, the chylific ventricle, that supplies the chalk. It keeps it separated from the food, either as original matter or as a derivative of the ammonium urate; it purges it of all foreign bodies, when the larval period comes to an end, and holds it in reserve until the time comes to disgorge it. This freestone factory causes me no astonishment: when the manufacturer undergoes his change, it serves for var-ious chemical works. Certain Oil-beetles, such as the Sitaris, locate in it the urate of ammonia, the refuse of the transformed organism; the Sphex, the Pelopaei, the Scoliae use it to manufacture the shellac wherewith the silk of the cocoon is var-nished. Further investigations will only swell the aggregate of the products of this obliging organ.

When the exit-way is prepared and the cell upholstered in velvet and closed with a threefold barricade, the industrious worm has concluded its task. It lays aside its tools, sheds its skin and becomes a nymph, a pupa, weakness personified, in swad-dling-clothes, on a soft couch. The head is always turned towards the door. This is a trifling detail in appearance; but it is everything in reality. To lie this way or that in the long cell is a matter of great indifference to the grub, which is very sup-ple, turning easily in its narrow lodging and adopting whatever position it pleas-es. The coming Capricorn will not enjoy the same privileges. Stiffly girt in his horn cuirass, he will not be able to turn from end to end; he will not even be capa-ble of bending, if some sudden wind should make the passage difficult. He must absolutely find the door in front of him, lest he perish in the casket. Should the grub forget this little formality, should it lie down to its nymphal sleep with its head at the back of the cell, the Capricorn is infallibly lost: his cradle becomes a hopeless dungeon.

But there is no fear of this danger: the knowledge of our bit of an intestine is too sound in things of the future for the grub to neglect the formality of keeping its head to the door. At the end of spring, the Capricorn, now in possession of his full strength, dreams of the joys of the sun, of the festivals of light. He wants to get out. What does he find before him? A heap of filings easily dispersed with his claws; next, a stone lid which he need not even break into fragments: it comes undone in one piece; it is removed from its frame with a few pushes of the forehead, a few tugs of the claws. In fact, I find the lid intact on the threshold of the abandoned cells. Last comes a second mass of woody remnants, as easy to disperse as the first. The road is now free: the Cerambyx has but to follow the spacious vestibule, which will lead him, without the possibility of mistake, to the exit. Should the window not be open, all that he has to do is to gnaw through a thin screen: an easy task; and behold him outside, his long antennae aquiver with excitement.

What have we learnt from him? Nothing, from him; much from his grub. This grub, so poor in sensory organs, gives us no little food for reflection with its pre-science. It knows that the coming Beetle will not be able to cut himself a road through the oak and it bethinks itself of opening one for him at its own risk and peril. It knows that the Cerambyx, in his stiff armour, will never be able to turn and make for the orifice of the cell; and it takes care to fall into its nymphal sleep with its head to the door. It knows how soft the pupa's flesh will be and upholsters the bedroom with velvet. It knows that the enemy is likely to break in during the slow work of the transformation and, to set a bulwark against his attacks, it stores a calcium pap inside its stomach. It knows the future with a clear vision, or, to be accurate, behaves as though it knew it. Whence did it derive the motives of its actions? Certainly not from the experience of the senses. What does it know of the outside world? Let us repeat, as much as a bit of an intestine can know. And this senseless creature fills us with amazement! I regret that the clever logician, instead of conceiving a statue smelling a rose, did not imagine it gifted with some instinct. How quickly he would have recognized that, quite apart from sense-impressions, the animal, including man, possesses certain psychological resources, certain inspirations that are innate and not acquired!

CHAPTER 5.

THE BURYING-BEETLES: THE BURIAL.

Beside the footpath in April lies the Mole, disembowelled by the peasant's spade; at the foot of the hedge the pitiless urchin has stoned to death the Lizard, who was about to don his green, pearl-embellished costume. The passer-by has thought it a meritorious deed to crush beneath his heel the chance-met Adder; and a gust of wind has thrown a tiny unfeathered bird from its nest. What will become of these little bodies and of so many other pitiful remnants of life? They will not long offend our sense of sight and smell. The sanitary officers of the fields are legion.

An eager freebooter, ready for any task, the Ant is the first to come hastening and begin, particle by particle, to dissect the corpse. Soon the odour of the corpse attracts the Fly, the genitrix of the odious maggot. At the same time, the flattened Silpha, the glistening, slow-trotting Horn-beetle, the Dermestes, powdered with snow upon the abdomen, and the slender Staphylinus, all, whence coming no one knows, hurry hither in squads, with never-wearied zeal, investigating, probing and draining the infection.

What a spectacle, in the spring, beneath a dead Mole! The horror of this laboratory is a beautiful sight for one who is able to observe and to meditate. Let us overcome our disgust; let us turn over the unclean refuse with our foot. What a swarming there is beneath it, what a tumult of busy workers! The Silphae, with wing-cases wide and dark, as though in mourning, fly distraught, hiding in the cracks in the soil; the Saprini, of polished ebony which mirrors the sunlight, jog hastily off, deserting their workshop; the Dermestes, of whom one wears a fawn-coloured tippet, spotted with white, seek to fly away, but, tipsy with their putrid nectar, tumble over and reveal the immaculate whiteness of their bellies, which forms a violent contrast with the gloom of the rest of their attire.

What were they doing there, all these feverish workers? They were making a clearance of death on behalf of life. Transcendent alchemists, they were transforming that horrible putridity into a living and inoffensive product. They were draining the dangerous corpse to the point of rendering it as dry and sonorous as the remains of an old slipper hardened on the refuse-heap by the frosts of winter and the heats of summer. They were working their hardest to render the carrion innocuous.

Others will soon put in their appearance, smaller creatures and more patient, who will take over the relic and exploit it ligament by ligament, bone by bone, hair by hair, until the whole has been resumed by the treasury of life. All honour to these purifiers! Let us put back the Mole and go our way.

Some other victim of the agricultural labours of spring—a Shrew-mouse, Field-mouse, Mole, Frog, Adder, or Lizard—will provide us with the most vigorous and famous of these expurgators of the soil. This is the Burying-beetle, the Necrophorus, so different from the cadaveric mob in dress and habits. In honour of his exalted functions he exhales an odour of musk; he bears a red tuft at the tip of his antennae; his breast is covered with nankeen; and across his wing-cases he wears a double, scalloped scarf of vermilion. An elegant, almost sumptuous costume, very superior to that of the others, but yet lugubrious, as befits your undertaker's man.

He is no anatomical dissector, cutting his subject open, carving its flesh with the scalpel of his mandibles; he is literally a gravedigger, a sexton. While the others—Silphae, Dermestes, Horn-beetles—gorge themselves with the exploited flesh, without, of course, forgetting the interests of the family, he, a frugal eater, hardly touches his booty on his own account. He buries it entire, on the spot, in a cellar where the thing, duly ripened, will form the diet of his larvae. He buries it in order to establish his progeny therein.

This hoarder of dead bodies, with his stiff and almost heavy movements, is astonishingly quick at storing away wreckage. In a shift of a few hours, a comparatively enormous animal—a Mole, for example—disappears, engulfed by the earth. The others leave the dried, emptied carcass to the air, the sport of the winds for months on end; he, treating it as a whole, makes a clean job of things at once. No visible trace of his work remains but a tiny hillock, a burial-mound, a tumulus.

With his expeditious method, the Necrophorus is the first of the little purifiers of the fields. He is also one of the most celebrated of insects in respect of his psychical capacities. This undertaker is endowed, they say, with intellectual faculties approaching to reason, such as are not possessed by the most gifted of the Bees and Wasps, the collectors of honey or game. He is honoured by the two following anecdotes, which I quote from Lacordaire's "Introduction to Entomology," the only general treatise at my disposal:

"Clairville," says the author, "records that he saw a Necrophorus vespillo, who, wishing to bury a dead Mouse and finding the soil on which the body lay too hard, proceeded to dig a hole at some distance in soil more easily displaced. This operation completed, he attempted to bury the Mouse in this cavity, but, not succeeding, he flew away, returning a few moments later accompanied by four of his fellows, who assisted him to move the Mouse and bury it."

In such actions, Lacordaire adds, we cannot refuse to admit the intervention of reason.

"The following case," he continues, "recorded by Gledditsch, has also every indication of the intervention of reason. One of his friends, wishing to desiccate a Frog, placed it on the top of a stick thrust into the ground, in order to make sure that the Necrophori should not come and carry it off. But this precaution was of no effect; the insects, being unable to reach the Frog, dug under the stick and, having caused it to fall, buried it as well as the body." ("Suites a Buffon. Introduction a l'entomologie" volume 2 pages 460-61.—Author's Note.)

To grant, in the intellect of the insect, a lucid understanding of the relations between cause and effect, between the end and the means, is an affirmation of serious import. I know of scarcely any better adapted to the philosophical brutalities of my time. But are these two little stories really true? Do they involve the consequences deduced from them? Are not those who accept them as reliable testimony a little over-simple?

To be sure, simplicity is needed in entomology. Without a good dose of this quality, a mental defect in the eyes of practical folk, who would busy himself with the lesser creatures? Yes, let us be simple, without being childishly credulous. Before making insects reason, let us reason a little ourselves; let us, above all, consult the experimental test. A fact gathered at hazard, without criticism, cannot establish a law.

I do not propose, O valiant grave-diggers, to belittle your merits; such is far from being my intention. I have that in my notes, on the other hand, which will do you

more honour than the case of the gibbet and the Frog; I have gleaned, for your benefit, examples of prowess which will shed a new lustre upon your reputation.

No, my intention is not to lessen your renown. However, it is not the business of
impartial history to maintain a given thesis; it follows whither the facts lead it. I
wish simply to question you upon the power of logic attributed to you. Do you
or do you not enjoy gleams of reason? Have you within you the humble germ of
human thought? That is the problem before us.

To solve it we will not rely upon the accidents which good fortune may now and
again procure for us. We must employ the breeding-cage, which will permit of
assiduous visits, continued inquiry and a variety of artifices. But how populate the
cage? The land of the olive-tree is not rich in Necrophori. To my knowledge it
possesses only a single species, N. vestigator (Hersch.); and even this rival of the
grave-diggers of the north is pretty scarce. The discovery of three or four in the
course of the spring was as much as my searches yielded in the old days. This time,
if I do not resort to the ruses of the trapper, I shall obtain them in no greater
numbers; whereas I stand in need of at least a dozen.

These ruses are very simple. To go in search of the layer-out of bodies, who exists
only here and there in the country-side, would be almost always waste of time;
the favourable month, April, would elapse before my cage was suitably populat-
ed. To run after him is to trust too much to accident; so we will make him come
to us by scattering in the orchard an abundant collection of dead Moles. To this
carrion, ripened by the sun, the insect will not fail to hasten from the various
points of the horizon, so accomplished is he in the detection of such a delicacy.

I make an arrangement with a gardener in the neighbourhood, who, two or three
times a week, supplements the penury of my acre and a half of stony ground, providing me with vegetables raised in a better soil. I explain to him my urgent need
of Moles, an indefinite number of moles. Battling daily with trap and spade
against the importunate excavator who uproots his crops, he is in a better position than any one else to procure for me that which I regard for the moment as
more precious than his bunches of asparagus or his white-heart cabbages.

The worthy man at first laughs at my request, being greatly surprised by the
importance which I attribute to the abhorrent creature, the Darboun; but at last
he consents, not without a suspicion at the back of his mind that I am going to
make myself a wonderful flannel-lined waist-coat with the soft, velvety skins of
the Moles, something good for pains in the back. Very well. We settle the matter.
The essential thing is that the Darbouns shall reach me.

They reach me punctually, by twos, by threes, by fours, packed in a few cabbage-leaves, at the bottom of the gardener's basket. The worthy man who lent himself with such good grace to my strange requirements will never guess how much comparative psychology will owe him! In a few days I was the possessor of thirty Moles, which were scattered here and there, as they reached me, in bare portions of the orchard, amid the rosemary-bushes, the arbutus-trees, and the lavender-beds.

Now it only remained to wait and to examine, several times a day, the under-side of my little corpses, a disgusting task which any one would avoid who had not the sacred fire in his veins. Only little Paul, of all the household, lent me the aid of his nimble hand to seize the fugitives. I have already stated that the entomologist has need of simplicity of mind. In this important business of the Necrophori, my assistants were a child and an illiterate.

Little Paul's visits alternating with mine, we had not long to wait. The four winds of heaven bore forth in all directions the odour of the carrion; and the undertakers hurried up, so that the experiments, begun with four subjects, were continued with fourteen, a number not attained during the whole of my previous searches, which were unpremeditated and in which no bait was used as decoy. My trapper's ruse was completely successful.

Before I report the results obtained in the cage, let us for a moment stop to consider the normal conditions of the labours that fall to the lot of the Necrophori. The Beetle does not select his head of game, choosing one in proportion to his strength, as do the predatory Wasps; he accepts it as hazard presents it to him. Among his finds there are little creatures, such as the Shrew-mouse; animals of medium size, such as the Field-mouse; and enormous beasts, such as the Mole, the Sewer-rat and the Snake, any of which exceeds the powers of excavation of a single grave-digger. In the majority of cases transportation is impossible, so disproportioned is the burden to the motive-power. A slight displacement, caused by the effort of the insects' backs, is all that can possibly be effected.

Ammophilus and Cerceris, Sphex and Pompilus excavate their burrows wherever they please; they carry their prey thither on the wing, or, if too heavy, drag it afoot. The Necrophorus knows no such facilities in his task. Incapable of carrying the monstrous corpse, no matter where encountered, he is forced to dig the grave where the body lies.

This obligatory place of sepulture may be in stony soil; it may occupy this or that bare spot, or some other where the grass, especially the couch-grass, plunges into the ground its inextricable network of little cords. There is a great probability, too, that a bristle of stunted brambles may support the body at some inches from the soil. Slung by the labourers' spade, which has just broken his back, the Mole falls here, there, anywhere, at random; and where the body falls, no matter what the obstacles—provided they be not insurmountable—there the undertaker must utilize it.

The difficulties of inhumation are capable of such variety as causes us already to foresee that the Necrophorus cannot employ fixed methods in the accomplishment of his labours. Exposed to fortuitous hazards, he must be able to modify his tactics within the limits of his modest perceptions. To saw, to break, to disentangle, to lift, to shake, to displace: these are so many methods of procedure which are indispensable to the grave-digger in a predicament. Deprived of these resources, reduced to uniformity of method, the insect would be incapable of pursuing the calling which has fallen to its lot.

We see at once how imprudent it would be to draw conclusions from an isolated case in which rational coordination or premeditated intention might appear to intervene. Every instinctive action no doubt has its motive; but does the animal in the first place judge whether the action is opportune? Let us begin by a careful consideration of the creature's labours; let us support each piece of evidence by others; and then we shall be able to answer the question.

First of all, a word as to diet. A general scavenger, the Burying-beetle refuses nothing in the way of cadaveric putridity. All is good to his senses, feathered game or furry, provided that the burden do not exceed his strength. He exploits the batrachian or the reptile with no less animation. he accepts without hesitation extraordinary finds, probably unknown to his race, as witness a certain Gold-fish, a red Chinese Carp, whose body, placed in one of my cages, was instantly considered an excellent tit-bit and buried according to the rules. Nor is butcher's meat despised. A mutton-cutlet, a strip of beefsteak, in the right stage of maturity, disappeared beneath the soil, receiving the same attention as those which were lavished on the Mole or the Mouse. In short, the Necrophorus has no exclusive preferences; anything putrid he conveys underground.

The maintenance of his industry, therefore, presents no sort of difficulty. If one kind of game be lacking, some other—the first to hand—will very well replace it. Neither is there much trouble in establishing the site of his industry. A capacious dish-cover of wire gauze is sufficient, resting on an earthen pan filled to the brim with fresh, heaped sand. To obviate criminal attempts on the part of the Cats, whom the game would not fail to tempt, the cage is installed in a closed room with glazed windows, which in winter is the refuge of the plants and in summer an entomological laboratory.

Now to work. The Mole lies in the centre of the enclosure. The soil, easily shifted and homogeneous, realizes the best conditions for comfortable work. Four Necrophori, three males and a female, are there with the body. They remain invisible, hidden beneath the carcass, which from time to time seems to return to life, shaken from end to end by the backs of the workers. An observer not in the secret would be somewhat astonished to see the dead creature move. From time to time, one of the sextons, almost always a male, emerges and goes the rounds of the animal, which he explores, probing its velvet coat. He hurriedly returns, appears again, once more investigates and creeps back under the corpse.

The tremors become more pronounced; the carcass oscillates, while a cushion of sand, pushed outward from below, grows up all about it. The Mole, by reason of his own weight and the efforts of the grave-diggers, who are labouring at their task beneath him, gradually sinks, for lack of support, into the undermined soil.

Presently the sand which has been pushed outward quivers under the thrust of the invisible miners, slips into the pit and covers the interred Mole. It is a clandestine burial. The body seems to disappear of itself, as though engulfed by a fluid medium. For a long time yet, until the depth is regarded as sufficient, the body will continue to descend.

It is, when all is taken into account, a very simple operation. As the diggers, underneath the corpse, deepen the cavity into which it sinks, tugged and shaken by the sextons, the grave, without their intervention, fills of itself by the mere downfall of the shaken soil. Useful shovels at the tips of their claws, powerful backs, capable of creating a little earthquake: the diggers need nothing more for the practice of their profession. Let us add—for this is an essential point—the art of continually jerking and shaking the body, so as to pack it into a lesser volume and cause it to pass when passage is obstructed. We shall presently see that this art plays a part of the greatest importance in the industry of the Necrophori.

Although he has disappeared, the Mole is still far from having reached his destination. Let us leave the undertakers to complete their task. What they are now doing below ground is a continuation of what they did on the surface and would teach us nothing new. We will wait for two or three days.

The moment has come. Let us inform ourselves of what is happening down there. Let us visit the retting-vat. I shall invite no one to be present at the exhumation. Of those about me, only little Paul has the courage to assist me.

The Mole is a Mole no longer, but a greenish horror, putrid, hairless, shrunk into a round, greasy mass. The thing must have undergone careful manipulation to be thus condensed into a small volume, like a fowl in the hands of the cook, and, above all, to be so completely deprived of its fur. Is this culinary procedure undertaken in respect of the larvae, which might be incommoded by the fur? Or is it just a casual result, a mere loss of hair due to putridity? I am not certain. But it is always the case that these exhumations, from first to last, have revealed the furry game furless and the feathered game featherless, except for the tail-feathers and the pinion-feathers of the wings. Reptiles and fish, on the other hand, retain their scales.

Let us return to the unrecognizable thing which was once a Mole. The tit-bit lies in a spacious crypt, with firm walls, a regular workshop, worthy of being the bake-house of a Copris-beetle. Except for the fur, which is lying in scattered flocks, it is intact. The grave-diggers have not eaten into it; it is the patrimony of the sons, not the provision of the parents, who, in order to sustain themselves, levy at most a few mouthfuls of the ooze of putrid humours.

Beside the dish which they are kneading and protecting are two Necrophori; a couple, no more. Four collaborated in the burial. What has become of the other two, both males? I find them hidden in the soil, at a distance, almost at the surface.

This observation is not an isolated one. Whenever I am present at a burial undertaken by a squad in which the males, zealous one and all, predominate, I find presently, when the burial is completed, only one couple in the mortuary cellar. Having lent their assistance, the rest have discreetly retired.

These grave-diggers, in truth, are remarkable fathers. They have nothing of the happy-go-lucky paternal carelessness that is the general rule among insects, which plague and pester the mother for a moment with their attentions and thereupon leave her to care for the offspring! But those who in the other races are unemployed in this case labour valiantly, now in the interest of their own family, now for the sake of another's, without distinction. If a couple is in difficulties, helpers arrive, attracted by the odour of carrion; anxious to serve a lady, they creep under the body, work at it with back and claw, bury it and then go their ways, leaving the householders to their happiness.

For some time longer these latter manipulate the morsel in concert, stripping it of fur or feather, trussing it and allowing it to simmer to the taste of the larvae. When all is in order, the couple go forth, dissolving their partnership, and each, following his fancy, recommences elsewhere, even if only as a mere auxiliary.

Twice and no oftener hitherto have I found the father preoccupied by the future of his sons and labouring in order to leave them rich: it happens with certain Dung-beetles and with the Necrophori, who bury dead bodies. Scavengers and undertakers both have exemplary morals. Who would look for virtue in such a quarter?

What follows—the larval existence and the metamorphosis—is a secondary detail and, for that matter, familiar. It is a dry subject and I shall deal with it briefly. About the end of May, I exhume a Brown Rat, buried by the grave-diggers a fortnight earlier. Transformed into a black, sticky jelly, the horrible dish provides me with fifteen larvae, already, for the most part, of the normal size. A few adults, connections, assuredly, of the brood, are also stirring amid the infected mass. The period of hatching is over now; and food is plentiful. Having nothing else to do, the foster-parents have sat down to the feast with the nurselings.

The undertakers are quick at rearing a family. It is at most a fortnight since the Rat was laid in the earth; and here already is a vigorous population on the verge of the metamorphosis. Such precocity amazes me. It would seem as though the liquefaction of carrion, deadly to any other stomach, is in this case a food productive of especial energy, which stimulates the organism and accelerates its growth, so that the victuals may be consumed before its approaching conversion into mould. Living chemistry makes haste to outstrip the ultimate reactions of mineral chemistry.

White, naked, blind, possessing the habitual attributes of life in darkness, the larva, with its lanceolate outline, is slightly reminiscent of the grub of the Ground-beetle. The mandibles are black and powerful, making excellent scissors for dissection. The limbs are short, but capable of a quick, toddling gait. The segments of the abdomen are armoured on the upper surface with a narrow reddish plate, armed with four tiny spikes, whose office apparently is to furnish points of support when the larva quits the natal dwelling and dives into the soil, there to undergo the transformation. The thoracic segments are provided with wider plates, but unarmed.

The adults discovered in the company of their larval family, in this putridity that was a Rat, are all abominably verminous. So shiny and neat in their attire, when at work under the first Moles of April, the Necrophori, when June approaches, become odious to look upon. A layer of parasites envelops them; insinuating itself into the joints, it forms an almost continuous surface. The insect presents a misshapen appearance under this overcoat of vermin, which my hair-pencil can hardly brush aside. Driven off the belly, the horde make the tour of the sufferer and encamp on his back, refusing to relinquish their hold.

I recognize among them the Beetle's Gamasis, the Tick who so often soils the ventral amethyst of our Geotrupes. No; the prizes of life do not fall to the share of the useful. Necrophori and Geotrupes devote themselves to works of general salubrity; and these two corporations, so interesting in the accomplishment of their hygienic functions, so remarkable for their domestic morality, are given over to the vermin of poverty. Alas, of this discrepancy between the services rendered and the harshness of life there are many other examples outside the world of scavengers and undertakers!

The Burying-beetles display an exemplary domestic morality, but it does not persist until the end. During the first fortnight of June, the family being sufficiently provided for, the sextons strike work and my cages are deserted, so far as the surface is concerned, in spite of new arrivals of Mice and Sparrows. From time to time some grave-digger leaves the subsoil and comes crawling languidly in the fresh air.

Another rather curious fact now attracts my attention. All, as soon as they emerge from underground, are cripples, whose limbs have been amputated at the joints, some higher up, some lower down. I see one mutilated Beetle who has only one leg left entire. With this odd limb and the stumps of the others lamentably tattered, scaly with vermin, he rows himself, as it were, over the dusty surface. A comrade emerges, one better off for legs, who finishes the cripple and cleans out his abdomen. So my thirteen remaining Necrophori end their days, half-devoured by their companions, or at least shorn of several limbs. The pacific relations of the outset are succeeded by cannibalism.

History tells us that certain peoples, the Massagetae and others, used to kill their aged folk in order to spare them the miseries of senility. The fatal blow on the hoary skull was in their eyes an act of filial piety. The Necrophori have their share of these ancient barbarities. Full of days and henceforth useless, dragging out a weary existence, they mutually exterminate one another. Why prolong the agony of the impotent and the imbecile?

The Massagetae might invoke, as an excuse for their atrocious custom, a dearth of provisions, which is an evil counsellor; not so the Necrophori, for, thanks to my generosity, victuals are superabundant, both beneath the soil and on the surface. Famine plays no part in this slaughter. Here we have the aberration of exhaustion, the morbid fury of a life on the point of extinction. As is generally the case, work bestows a peaceable disposition on the grave-digger, while inaction inspires him with perverted tastes. Having no longer anything to do, he breaks his fellow's limbs, eats him up, heedless of being mutilated or eaten up himself. This is the ultimate deliverance of verminous old age.

CHAPTER 6.

THE BURYING-BEETLES: EXPERIMENTS.

Let us proceed to the rational prowess which has earned for the Necrophorus the better part of his renown and, to begin with, let us submit the case related by Clairville—that of the too hard soil and the call for assistance—to experimental test.

With this object in view, I pave the centre of the space beneath the cover, level with the soil, with a brick and sprinkle the latter with a thin layer of sand. This will be the soil in which digging is impracticable. All about it, for some distance and on the same level, spreads the loose soil, which is easy to dig.

In order to approximate to the conditions of the little story, I must have a Mouse; with a Mole, a heavy mass, the work of removal would perhaps present too much difficulty. To obtain the Mouse I place my friends and neighbours under requisition; they laugh at my whim but none the less proffer their traps. Yet, the moment a Mouse is needed, that very common animal becomes rare. Braving decorum in his speech, which follows the Latin of his ancestors, the Provençal says, but even more crudely than in my translation: "If you look for dung, the Asses become constipated!"

At last I possess the Mouse of my dreams! She comes to me from that refuge, furnished with a truss of straw, in which official charity gives the hospitality of a day to the beggar wandering over the face of the fertile earth; from that municipal hostel whence one invariably emerges verminous. O Réaumur, who used to invite

marquises to see your caterpillars change their skins, what would you have said of
a future disciple conversant with such wretchedness as this? Perhaps it is well that
we should not be ignorant of it, so that we may take compassion on the suffer-
ings of beasts.

The Mouse so greatly desired is mine. I place her upon the centre of the brick.
The grave-diggers under the wire cover are now seven in number, of whom three
are females. All have gone to earth: some are inactive, close to the surface; the rest
are busy in their crypts. The presence of the fresh corpse is promptly perceived.
About seven o'clock in the morning, three Necrophori hurry up, two males and
a female. They slip under the Mouse, who moves in jerks, a sign of the efforts of
the burying-party. An attempt is made to dig into the layer of sand which hides
the brick, so that a bank of sand accumulates about the body.

For a couple of hours the jerks continue without results. I profit by the circumstance to investigate the manner in which the work is performed. The bare brick
allows me to see what the excavated soil concealed from me. If it is necessary to
move the body, the Beetle turns over; with his six claws he grips the hair of the
dead animal, props himself upon his back and pushes, making a lever of his head
and the tip of his abdomen. If digging is required, he resumes the normal position. So, turn and turn about, the sexton strives, now with his claws in the air,
when it is a question of shifting the body or dragging it lower down; now with
his feet on the ground, when it is necessary to deepen the grave.

The point at which the Mouse lies is finally recognized as unassailable. A male
appears in the open. He explores the specimen, goes the round of it, scratches a
little at random. He goes back; and immediately the body rocks. Is he advising his
collaborators of what he has discovered? Is he arranging matters with a view to
their establishing themselves elsewhere, on propitious soil?

The facts are far from confirming this idea. When he shakes the body, the others
imitate him and push, but without combining their efforts in a given direction,
for, after advancing a little towards the edge of the brick, the burden goes back
again, returning to the point of departure. In the absence of any concerted understanding, their efforts of leverage are wasted. Nearly three hours are occupied by
oscillations which mutually annul one another. The Mouse does not cross the little sand-hill heaped about it by the rakes of the workers.

For the second time a male emerges and makes a round of exploration. A bore is
made in workable earth, close beside the brick. This is a trial excavation, to reveal
the nature of the soil; a narrow well, of no great depth, into which the insect

plunges to half its length. The well-sinker returns to the other workers, who arch their backs, and the load progresses a finger's-breadth towards the point recognized as favourable. Have they done the trick this time? No, for after a while the Mouse recoils. No progress towards a solution of the difficulty.

Now two males come out in search of information, each of his own accord. Instead of stopping at the point already sounded, a point most judiciously chosen, it seemed, on account of its proximity, which would save laborious transportation, they precipitately scour the whole area of the cage, sounding the soil on this side and on that and ploughing superficial furrows in it. They get as far from the brick as the limits of the enclosure permit.

They dig, by preference, against the base of the cover; here they make several borings, without any reason, so far as I can see, the bed of soil being everywhere equally assailable away from the brick; the first point sounded is abandoned for a second, which is rejected in its turn. A third and a fourth are tried; then another and yet another. At the sixth point the selection is made. In all these cases the excavation is by no means a grave destined to receive the Mouse, but a mere trial boring, of inconsiderable depth, its diameter being that of the digger's body.

A return is made to the Mouse, who suddenly quivers, oscillates, advances, recoils, first in one direction, then in another, until in the end the little hillock of sand is crossed. Now we are free of the brick and on excellent soil. Little by little the load advances. This is no cartage by a team hauling in the open, but a jerky displacement, the work of invisible levers. The body seems to move of its own accord.

This time, after so many hesitations, their efforts are concerted; at all events, the load reaches the region sounded far more rapidly than I expected. Then begins the burial, according to the usual method. It is one o'clock. The Necrophori have allowed the hour-hand of the clock to go half round the dial while verifying the condition of the surrounding spots and displacing the Mouse.

In this experiment it appears at the outset that the males play a major part in the affairs of the household. Better-equipped, perhaps, than their mates, they make investigations when a difficulty occurs; they inspect the soil, recognize whence the check arises and choose the point at which the grave shall be made. In the lengthy experiment of the brick, the two males alone explored the surroundings and set to work to solve the difficulty. Confiding in their assistance, the female, motionless beneath the Mouse, awaited the result of their investigations. The tests which are to follow will confirm the merits of these valiant auxiliaries.

In the second place, the point where the Mouse lay being recognized as presenting an insurmountable resistance, there was no grave dug in advance, a little farther off, in the light soil. All attempts were limited, I repeat, to shallow soundings
which informed the insect of the possibility of inhumation.

It is absolute nonsense to speak of their first preparing the grave to which the
body will afterwards be carted. To excavate the soil, our grave-diggers must feel
the weight of their dead on their backs. They work only when stimulated by the
contact of its fur. Never, never in this world do they venture to dig a grave unless
the body to be buried already occupies the site of the cavity. This is absolutely
confirmed by my two and a half months and more of daily observations.

The rest of Clairville's anecdote bears examination no better. We are told that the
Necrophorus in difficulties goes in search of assistance and returns with companions who assist him to bury the Mouse. This, in another form, is the edifying
story of the Sacred Beetle whose pellet had rolled into a rut. powerless to withdraw his treasure from the gulf, the wily Dung-beetle called together three or four
of his neighbours, who benevolently recovered the pellet, returning to their
labours after the work of salvage.

The exploit—so ill-interpreted—of the thieving pill-roller sets me on my guard
against that of the undertaker. Shall I be too exigent if I enquire what precautions
the observer adopted to recognize the owner of the Mouse on his return, when he
reappears, as we are told, with four assistants? What sign denotes that one of the
five who was able, in so rational a manner, to appeal for help? Can one even be
sure that the one to disappear returns and forms one of the band? There is nothing to indicate it; and this was the essential point which a sterling observer was
bound not to neglect. Were they not rather five chance Necrophori who, guided
by the smell, without any previous understanding, hastened to the abandoned
Mouse to exploit her on their own account? I incline to this opinion, the most
likely of all in the absence of exact information.

Probability becomes certainty if we submit the case to the verification of experiment. The test with the brick already gives us some information. For six hours my
three specimens exhausted themselves in efforts before they got to the length of
removing their booty and placing it on practicable soil. In this long and heavy
task helpful neighbours would have been anything but unwelcome. Four other
Necrophori, buried here and there under a little sand, comrades and acquaintances, helpers of the day before, were occupying the same cage; and not one of
those concerned thought of summoning them to give assistance. Despite their
extreme embarrassment, the owners of the Mouse accomplished their task to the
end, without the least help, though this could have been so easily requisitioned.

Being three, one might say, they considered themselves sufficiently strong; they needed no one else to lend them a hand. The objection does not hold good. On many occasions and under conditions even more difficult than those presented by a stony soil, I have again and again seen isolated Necrophori exhausting themselves in striving against my artifices; yet not once did they leave their work to recruit helpers. Collaborators, it is true, did often arrive, but they were convoked by their sense of smell, not by the first possessor. They were fortuitous helpers; they were never called in. They were welcomed without disagreement, but also without gratitude. They were not summoned; they were tolerated. In the glazed shelter where I keep the cage I happened to catch one of these chance assistants in the act. Passing that way in the night and scenting dead flesh, he had entered where none of his kind had yet penetrated of his own free will. I surprised him on the wire-gauze dome of the cover. If the wire had not prevented him, he would have set to work incontinently, in company with the rest. Had my captives invited him? Assuredly not. He had hastened thither attracted by the odour of the Mole, heedless of the efforts of others. So it was with those whose obliging assistance is extolled. I repeat, in respect of their imaginary prowess, what I have said elsewhere of that of the Sacred Beetles: the story is a childish one, worthy of ranking with any fairy-tale written for the amusement of the simple.

A hard soil, necessitating the removal of the body, is not the only difficulty familiar to the Necrophori. Often, perhaps more often than not, the ground is covered with grass, above all with couch-grass, whose tenacious rootlets form an inextricable network below the surface. To dig in the interstices is possible, but to drag the dead animal through them is another matter: the meshes of the net are too close to give it passage. Will the grave-digger find himself reduced to impotence by such an impediment, which must be an extremely common one? That could not be.

Exposed to this or that habitual obstacle in the exercise of his calling, the animal is always equipped accordingly; otherwise his profession would be impracticable. No end is attained without the necessary means and aptitudes. Besides that of the excavator, the Necrophorus certainly possesses another art: the art of breaking the cables, the roots, the stolons, the slender rhizomes which check the body's descent into the grave. To the work of the shovel and the pick must be added that of the shears. All this is perfectly logical and may be foreseen with complete lucidity. Nevertheless, let us invoke experiment, the best of witnesses.

I borrow from the kitchen-range an iron trivet whose legs will supply a solid foundation for the engine which I am devising. This is a coarse network of strips of raphia, a fairly accurate imitation of the network of couch-grass roots. The very irregular meshes are nowhere wide enough to admit of the passage of the creature to be buried, which in this case is a Mole. The trivet is planted with its three feet in the soil of the cage; its top is level with the surface of the soil. A little sand conceals the meshes. The Mole is placed in the centre; and my squad of sextons is let loose upon the body.

Without a hitch the burial is accomplished in the course of an afternoon. The hammock of raphia, almost equivalent to the natural network of couch-grass turf, scarcely disturbs the process of inhumation. Matters do not go forward quite so quickly; and that is all. No attempt is made to shift the Mole, who sinks into the ground where he lies. The operation completed, I remove the trivet. The network is broken at the spot where the corpse lay. A few strips have been gnawed through; a small number, only so many as were strictly necessary to permit the passage of the body.

Well done, my undertakers! I expected no less of your savoir-faire. You have foiled the artifices of the experimenter by employing your resources against natural obstacles. With mandibles for shears, you have patiently cut my threads as you would have gnawed the cordage of the grass-roots. This is meritorious, if not deserving of exceptional glorification. The most limited of the insects which work in earth would have done as much if subjected to similar conditions.

Let us ascend a stage in the series of difficulties. The Mole is now fixed with a lashing of raphia fore and aft to a light horizontal cross-bar which rests on two firmly-planted forks. It is like a joint of venison on a spit, though rather oddly fastened. The dead animal touches the ground throughout the length of its body.

The Necrophori disappear under the corpse, and, feeling the contact of its fur, begin to dig. The grave grows deeper and an empty space appears, but the coveted object does not descend, retained as it is by the cross-bar which the two forks keep in place. The digging slackens, the hesitations become prolonged.

However, one of the grave-diggers ascends to the surface, wanders over the Mole, inspects him and ends by perceiving the hinder strap. Tenaciously he gnaws and ravels it. I hear the click of the shears that completes the rupture. Crack! The thing is done. Dragged down by his own weight, the Mole sinks into the grave, but slantwise, with his head still outside, kept in place by the second ligature.

The Beetles proceed to the burial of the hinder part of the Mole; they twitch and jerk it now in this direction, now in that. Nothing comes of it; the thing refuses to give. A fresh sortie is made by one of them to discover what is happening overhead. The second ligature is perceived, is severed in turn, and henceforth the work proceeds as well as could be desired.

My compliments, perspicacious cable-cutters! But I must not exaggerate. The lashings of the Mole were for you the little cords with which you are so familiar in turfy soil. You have severed them, as well as the hammock of the previous experiment, just as you sever with the blades of your shears any natural filament which stretches across your catacombs. It is, in your calling, an indispensable knack. If you had had to learn it by experience, to think it out before practising it, your race would have disappeared, killed by the hesitations of its apprenticeship, for the spots fertile in Moles, Frogs, Lizards and other victuals to your taste are usually grass-covered.

You are capable of far better things yet; but, before proceeding to these, let us examine the case when the ground bristles with slender brushwood, which holds the corpse at a short distance from the ground. Will the find thus suspended by the hazard of its fall remain unemployed? Will the Necrophori pass on, indifferent to the superb tit-bit which they see and smell a few inches above their heads, or will they make it descend from its gibbet?

Game does not abound to such a point that it can be disdained if a few efforts will obtain it. Before I see the thing happen I am persuaded that it will fall, that the Necrophori, often confronted by the difficulties of a body which is not lying on the soil, must possess the instinct to shake it to the ground. The fortuitous support of a few bits of stubble, of a few interlaced brambles, a thing so common in the fields, should not be able to baffle them. The overthrow of the suspended body, if placed too high, should certainly form part of their instinctive methods. For the rest, let us watch them at work.

I plant in the sand of the cage a meagre tuft of thyme. The shrub is at most some four inches in height. In the branches I place a Mouse, entangling the tail, the paws and the neck among the twigs in order to increase the difficulty. The population of the cage now consists of fourteen Necrophori and will remain the same until the close of my investigations. Of course they do not all take part simultaneously in the day's work; the majority remain underground, somnolent, or occupied in setting their cellars in order. Sometimes only one, often two, three or four, rarely more, busy themselves with the dead creature which I offer them. To-day two hasten to the Mouse, who is soon perceived overhead in the tuft of thyme.

They gain the summit of the plant by way of the wire trellis of the cage. Here are repeated, with increased hesitation, due to the inconvenient nature of the support, the tactics employed to remove the body when the soil is unfavourable. The insect props itself against a branch, thrusting alternately with back and claws, jerking and shaking vigorously until the point where at it is working is freed from its fetters. In one brief shift, by dint of heaving their backs, the two collaborators extricate the body from the entanglement of twigs. Yet another shake; and the Mouse is down. The burial follows.

There is nothing new in this experiment; the find has been dealt with just as though it lay upon soil unsuitable for burial. The fall is the result of an attempt to transport the load.

The time has come to set up the Frog's gibbet celebrated by Gledditsch. The batrachian is not indispensable; a Mole will serve as well or even better. With a ligament of raphia I fix him, by his hind-legs, to a twig which I plant vertically in the ground, inserting it to no great depth. The creature hangs plumb against the gibbet, its head and shoulders making ample contact with the soil.

The gravediggers set to work beneath the part which lies upon the ground, at the very foot of the stake; they dig a funnel-shaped hole, into which the muzzle, the head and the neck of the mole sink little by little. The gibbet becomes uprooted as they sink and eventually falls, dragged over by the weight of its heavy burden. I am assisting at the spectacle of the overturned stake, one of the most astonishing examples of rational accomplishment which has ever been recorded to the credit of the insect.

This, for one who is considering the problem of instinct, is an exciting moment. But let us beware of forming conclusions as yet; we might be in too great a hurry. Let us ask ourselves first whether the fall of the stake was intentional or fortuitous. Did the Necrophori lay it bare with the express intention of causing it to fall? Or did they, on the contrary, dig at its base solely in order to bury that part of the mole which lay on the ground? that is the question, which, for the rest, is very easy to answer.

The experiment is repeated; but this time the gibbet is slanting and the Mole, hanging in a vertical position, touches the ground at a couple of inches from the base of the gibbet. Under these conditions absolutely no attempt is made to overthrow the latter. Not the least scrape of a claw is delivered at the foot of the gibbet. The entire work of excavation is accomplished at a distance, under the body, whose shoulders are lying on the ground. There—and there only—a hole is dug to receive the free portion of the body, the part accessible to the sextons.

A difference of an inch in the position of the suspended animal annihilates the famous legend. Even so, many a time, the most elementary sieve, handled with a little logic, is enough to winnow the confused mass of affirmations and to release the good grain of truth.

Yet another shake of the sieve. The gibbet is oblique or vertical indifferently; but the Mole, always fixed by a hinder limb to the top of the twig, does not touch the soil; he hangs a few fingers'-breadths from the ground, out of the sextons' reach.

What will the latter do? Will they scrape at the foot of the gibbet in order to overturn it? By no means; and the ingenuous observer who looked for such tactics would be greatly disappointed. No attention is paid to the base of the support. It is not vouchsafed even a stroke of the rake. Nothing is done to overturn it, nothing, absolutely nothing! It is by other methods that the Burying-beetles obtain the Mole.

These decisive experiments, repeated under many different forms, prove that never, never in this world do the Necrophori dig, or even give a superficial scrape, at the foot of the gallows, unless the hanging body touch the ground at that point. And, in the latter case, if the twig should happen to fall, its fall is in nowise an intentional result, but a mere fortuitous effect of the burial already commenced.

What, then, did the owner of the Frog of whom Gledditsch tells us really see? If his stick was overturned, the body placed to dry beyond the assaults of the Necrophori must certainly have touched the soil: a strange precaution against robbers and the damp! We may fittingly attribute more foresight to the preparer of dried Frogs and allow him to hang the creature some inches from the ground. In this case all my experiments emphatically assert that the fall of the stake undermined by the sextons is a pure matter of imagination.

Yet another of the fine arguments in favour of the reasoning power of animals flies from the light of investigation and founders in the slough of error! I admire your simple faith, you masters who take seriously the statements of chance-met observers, richer in imagination than in veracity; I admire your credulous zeal, when, without criticism, you build up your theories on such absurdities.

Let us proceed. The stake is henceforth planted vertically, but the body hanging on it does not reach the base: a condition which suffices to ensure that there is never any digging at this point. I make use of a Mouse, who, by reason of her trifling weight, will lend herself better to the insect's manoeuvres. The dead body is fixed by the hind-legs to the top of the stake with a ligature of raphia. It hangs plumb, in contact with the stick.

Very soon two Necrophori have discovered the tit-bit. They climb up the miniature mast; they explore the body, dividing its fur by thrusts of the head. It is recognized to be an excellent find. So to work. Here we have again, but under far
more difficult conditions, the tactics employed when it was necessary to displace
the unfavourably situated body: the two collaborators slip between the Mouse and
the stake, when, taking a grip of the latter and exerting a leverage with their backs,
they jerk and shake the body, which oscillates, twirls about, swings away from the
stake and relapses. All the morning is passed in vain attempts, interrupted by
explorations on the animal's body.

In the afternoon the cause of the check is at last recognized; not very clearly, for
in the first place the two obstinate riflers of the gallows attack the hind-legs of the
Mouse, a little below the ligature. They strip them bare, flay them and cut away
the flesh about the heel. They have reached the bone, when one of them finds the
raphia beneath his mandibles. This, to him, is a familiar thing, representing the
gramineous fibre so frequent in the case of burial in grass-covered soil.
Tenaciously the shears gnaw at the bond; the vegetable fetter is severed and the
Mouse falls, to be buried a little later.

If it were isolated, this severance of the suspending tie would be a magnificent
performance; but considered in connection with the sum of the Beetle's customary labours it loses all far-reaching significance. Before attacking the ligature,
which was not concealed in any way, the insect exerted itself for a whole morning
in shaking the body, its usual method. Finally, finding the cord, it severed it, as it
would have severed a ligament of couch-grass encountered underground.

Under the conditions devised for the Beetle, the use of the shears is the indispensable complement of the use of the shovel; and the modicum of discernment at
his disposal is enough to inform him when the blades of his shears will be useful.
He cuts what embarrasses him with no more exercise of reason than he displays
when placing the corpse underground. So little does he grasp the connection
between cause and effect that he strives to break the bone of the leg before gnawing at the bast which is knotted close beside him. The difficult task is attacked
before the extremely simple.

Difficult, yes, but not impossible, provided that the Mouse be young. I begin
again with a ligature of iron wire, on which the shears of the insect can obtain no
purchase, and a tender Mouselet, half the size of an adult. This time a tibia is
gnawed through, cut in two by the Beetle's mandibles near the spring of the heel.
The detached member leaves plenty of space for the other, which readily slips
from the metallic band; and the little body falls to the ground.

But, if the bone be too hard, if the body suspended be that of a Mole, an adult Mouse, or a Sparrow, the wire ligament opposes an insurmountable obstacle to the attempts of the Necrophori, who, for nearly a week, work at the hanging body, partly stripping it of fur or feather and dishevelling it until it forms a lamentable object, and at last abandon it, when desiccation sets in. A last resource, however, remains, one as rational as infallible. It is to overthrow the stake. Of course, not one dreams of doing so.

For the last time let us change our artifices. The top of the gibbet consists of a little fork, with the prongs widely opened and measuring barely two-fifths of an inch in length. With a thread of hemp, less easily attacked than a strip of raphia, I bind together, a little above the heels, the hind-legs of an adult Mouse; and between the legs I slip one of the prongs of the fork. To make the body fall it is enough to slide it a little way upwards; it is like a young Rabbit hanging in the front of a poulterer's shop.

Five Necrophori come to inspect my preparations. After a great deal of futile shaking, the tibiae are attacked. This, it seems, is the method usually employed when the body is retained by one of its limbs in some narrow fork of a low-growing plant. While trying to saw through the bone—a heavy job this time—one of the workers slips between the shackled limbs. So situated, he feels against his back the furry touch of the Mouse. Nothing more is needed to arouse his propensity to thrust with his back. With a few heaves of the lever the thing is done; the Mouse rises a little, slides over the supporting peg and falls to the ground.

Is this manoeuvre really thought out? Has the insect indeed perceived, by the light of a flash of reason, that in order to make the tit-bit fall it was necessary to unhook it by sliding it along the peg? Has it really perceived the mechanism of suspension? I know some persons—indeed, I know many—who, in the presence of this magnificent result, would be satisfied without further investigation.

More difficult to convince, I modify the experiment before drawing a conclusion. I suspect that the Necrophorus, without any prevision of the consequences of his action, heaved his back simply because he felt the legs of the creature above him. With the system of suspension adopted, the push of the back, employed in all cases of difficulty, was brought to bear first upon the point of support; and the fall resulted from this happy coincidence. That point, which has to be slipped along the peg in order to unhook the object, ought really to be situated at a short distance from the Mouse, so that the Necrophori shall no longer feel her directly against their backs when they push.

A piece of wire binds together now the tarsi of a Sparrow, now the heels of a Mouse and is bent, at a distance of three-quarters of an inch or so, into a little ring, which slips very loosely over one of the prongs of the fork, a short, almost horizontal prong. To make the hanging body fall, the slightest thrust upon this ring is sufficient; and, owing to its projection from the peg, it lends itself excellently to the insect's methods. In short, the arrangement is the same as it was just now, with this difference, that the point of support is at a short distance from the suspended animal.

My trick, simple though it be, is fully successful. For a long time the body is repeatedly shaken, but in vain; the tibiae or tarsi, unduly hard, refuse to yield to the patient saw. Sparrows and Mice grow dry and shrivelled, unused, upon the gibbet. Sooner in one case, later in another, my Necrophori abandon the insoluble problem in mechanics: to push, ever so little, the movable support and so to unhook the coveted carcass.

Curious reasoners, in faith! If they had had, but now, a lucid idea of the mutual relations between the shackled limbs and the suspending peg; if they had made the Mouse fall by a reasoned manoeuvre, whence comes it that the present artifice, no less simple than the first, is to them an insurmountable obstacle? For days and days they work on the body, examine it from head to foot, without becoming aware of the movable support, the cause of their misadventure. In vain do I prolong my watch; never do I see a single one of them push it with his foot or butt it with his head.

Their defeat is not due to lack of strength. Like the Geotrupes, they are vigorous excavators. Grasped in the closed hand, they insinuate themselves through the interstices of the fingers and plough up your skin in a fashion to make you very quickly loose your hold. With his head, a robust ploughshare, the Beetle might very easily push the ring off its short support. He is not able to do so because he does not think of it; he does not think of it because he is devoid of the faculty attributed to him, in order to support its thesis, by the dangerous prodigality of transformism.

Divine reason, sun of the intellect, what a clumsy slap in thy august countenance, when the glorifiers of the animal degrade thee with such dullness!

Let us now examine under another aspect the mental obscurity of the Necrophori. My captives are not so satisfied with their sumptuous lodging that they do not seek to escape, especially when there is a dearth of labour, that sovran consoler of the afflicted, man or beast. Internment within the wire cover palls upon them. So, the Mole buried and all in order in the cellar, they stray uneasily over the wire-gauze of the dome; they clamber up, descend, ascend again and take to flight, a flight which instantly becomes a fall, owing to collision with the wire grating. They pick themselves up and begin again. The sky is superb; the weather is hot, calm and propitious for those in search of the Lizard crushed beside the footpath. Perhaps the effluvia of the gamy tit-bit have reached them, coming from afar, imperceptible to any other sense than that of the Sexton-beetles. So my Necrophori are fain to go their ways.

Can they? Nothing would be easier if a glimmer of reason were to aid them. Through the wire network, over which they have so often strayed, they have seen, outside, the free soil, the promised land which they long to reach. A hundred times if once have they dug at the foot of the rampart. There, in vertical wells, they take up their station, drowsing whole days on end while unemployed. If I give them a fresh Mole, they emerge from their retreat by the entrance corridor and come to hide themselves beneath the belly of the beast. The burial over, they return, one here, one there, to the confines of the enclosure and disappear beneath the soil.

Well, in two and a half months of captivity, despite long stays at the base of the trellis, at a depth of three-quarters of an inch beneath the surface, it is rare indeed for a Necrophorus to succeed in circumventing the obstacle, to prolong his excavation beneath the barrier, to make an elbow in it and to bring it out on the other side, a trifling task for these vigorous creatures. Of fourteen only one succeeded in escaping.

A chance deliverance and not premeditated; for, if the happy event had been the result of a mental combination, the other prisoners, practically his equals in powers of perception, would all, from first to last, discover by rational means the elbowed path leading to the outer world; and the cage would promptly be deserted. The failure of the great majority proves that the single fugitive was simply digging at random. Circumstances favoured him; and that is all. Do not let us make it a merit that he succeeded where all the others failed.

Let us also beware of attributing to the Necrophori an understanding more limited than is usual in entomological psychology. I find the ineptness of the undertaker in all the insects reared under the wire cover, on the bed of sand into which the rim of the dome sinks a little way. With very rare exceptions, fortuitous accidents, no insect has thought of circumventing the barrier by way of the base; none has succeeded in gaining the exterior by means of a slanting tunnel, not even though it were a miner by profession, as are the Dung-beetles par excellence. Captives under the wire dome, but desirous of escape, Sacred Beetles, Geotrupes, Copres, Gymnopleuri, Sisyphi, all see about them the freedom of space, the joys of the open sunlight; and not one thinks of going round under the rampart, a front which would present no difficulty to their pick-axes.

Even in the higher ranks of animality, examples of similar mental obfuscation are not lacking. Audubon relates how, in his days, the wild Turkeys were caught in North America.

In a clearing known to be frequented by these birds, a great cage was constructed with stakes driven into the ground. In the centre of the enclosure opened a short tunnel, which dipped under the palisade and returned to the surface outside the cage by a gentle slope, which was open to the sky. The central opening, large enough to give a bird free passage, occupied only a portion of the enclosure, leaving around it, against the circle of stakes, a wide unbroken zone. A few handfuls of maize were scattered in the interior of the trap, as well as round about it, and in particular along the sloping path, which passed under a sort of bridge and led to the centre of the contrivance. In short, the Turkey-trap presented an ever-open door. The bird found it in order to enter, but did not think of looking for it in order to return by it.

According to the famous American ornithologist, the Turkeys, lured by the grains of maize, descended the insidious slope, entered the short underground passage and beheld, at the end of it, plunder and the light. A few steps farther and the gluttons emerged, one by one, from beneath the bridge. They distributed themselves about the enclosure. The maize was abundant; and the Turkeys' crops grew swollen.

When all was gathered, the band wished to retreat, but not one of the prisoners paid any attention to the central hole by which he had arrived. Gobbling uneasily, they passed again and again across the bridge whose arch was yawning beside them; they circled round against the palisade, treading a hundred times in their own footprints; they thrust their necks, with their crimson wattles, through the bars; and there, with beaks in the open air, they remained until they were exhausted.

Remember, inept fowl, the occurrences of a little while ago; think of the tunnel which led you hither! If there be in that poor brain of yours an atom of capacity, put two ideas together and remind yourself that the passage by which you entered is there and open for your escape! You will do nothing of the kind. The light, an irresistible attraction, holds you subjugated against the palisade; and the shadow of the yawning pit, which has but lately permitted you to enter and will quite as readily permit of your exit, leaves you indifferent. To recognize the use of this opening you would have to reflect a little, to evolve the past; but this tiny retrospective calculation is beyond your powers. So the trapper, returning a few days later, will find a rich booty, the entire flock imprisoned!

Of poor intellectual repute, does the Turkey deserve his name for stupidity? He does not appear to be more limited than another. Audubon depicts him as endowed with certain useful ruses, in particular when he has to baffle the attacks of his nocturnal enemy, the Virginian Owl. As for his actions in the snare with the underground passage, any other bird, impassioned of the light, would do the same.

Under rather more difficult conditions, the Necrophorus repeats the ineptness of the Turkey. When he wishes to return to the open daylight, after resting in a short burrow against the rim of the wire cover, the Beetle, seeing a little light filtering down through the loose soil, reascends by the path of entry, incapable of telling himself that it would suffice to prolong the tunnel as far in the opposite direction for him to reach the outer world beyond the wall and gain his freedom. Here again is one in whom we shall seek in vain for any indication of reflection. Like the rest, in spite of his legendary renown, he has no guide but the unconscious promptings of instinct.

CHAPTER 7.

THE BLUEBOTTLE.

To purge the earth of death's impurities and cause deceased animal matter to be once more numbered among the treasures of life there are hosts of sausage-queens, including, in our part of the world, the Bluebottle (Calliphora vomitaria, Lin.) and the Grey Flesh-fly (Sarcophaga carnaria, Lin.) Every one knows the first, the big, dark-blue Fly who, after effecting her designs in the ill-watched meat-safe, settles on our window-panes and keeps up a solemn buzzing, anxious to be off in the sun and ripen a fresh emission of germs. How does she lay her eggs, the origin of the loathsome maggot that battens poisonously on our provisions whether of game or butcher's meat? What are her stratagems and how can we foil them? This is what I propose to investigate.

The Bluebottle frequents our homes during autumn and a part of winter, until the cold becomes severe; but her appearance in the fields dates back much earlier. On the first fine day in February, we shall see her warming herself, chillily, against the sunny walls. In April, I notice her in considerable numbers on the laurustinus. It is here that she seems to pair, while sipping the sugary exudations of the small white flowers. The whole of the summer season is spent out of doors, in brief flights from one refreshment-bar to the next. When autumn comes, with its game, she makes her way into our houses and remains until the hard frosts.

This suits my stay-at-home habits and especially my legs, which are bending under the weight of years. I need not run after the subjects of my present study; they call on me. Besides, I have vigilant assistants. The household knows of my plans. One and all bring me, in a little screw of paper, the noisy visitor just captured against the panes.

Thus do I fill my vivarium, which consists of a large, bell-shaped cage of wire-gauze, standing in an earthenware pan full of sand. A mug containing honey is the dining-room of the establishment. Here the captives come to recruit themselves in their hours of leisure. To occupy their maternal cares, I employ small birds—Chaffinches, Linnets, Sparrows—brought down, in the enclosure, by my son's gun.

I have just served up a Linnet shot two days ago. I next place in the cage a Bluebottle, one only, to avoid confusion. Her fat belly proclaims the advent of laying-time. An hour later, when the excitement of being put in prison is allayed, my captive is in labour. With eager, jerky steps, she explores the morsel of game, goes from the head to the tail, returns from the tail to the head, repeats the action several times and at last settles near an eye, a dimmed eye sunk into its socket.

The ovipositor bends at a right angle and dives into the junction of the beak, straight down to the root. Then the eggs are emitted for nearly half an hour. The layer, utterly absorbed in her serious business, remains stationary and impassive and is easily observed through my lens. A movement on my part would doubtless scare her; but my restful presence gives her no anxiety. I am nothing to her.

The discharge does not go on continuously until the ovaries are exhausted; it is intermittent and performed in so many packets. Several times over, the Fly leaves the bird's beak and comes to take a rest upon the wire-gauze, where she brushes her hind-legs one against the other. In particular, before using it again, she cleans, smooths and polishes her laying-tool, the probe that places the eggs. Then, feeling her womb still teeming, she returns to the same spot at the joint of the beak. The delivery is resumed, to cease presently and then begin anew. A couple of hours are thus spent in alternate standing near the eye and resting on the wire-gauze.

At last it is over. The Fly does not go back to the bird, a proof that her ovaries are exhausted. The next day she is dead. The eggs are dabbed in a continuous layer, at the entrance to the throat, at the root of the tongue, on the membrane of the palate. Their number appears considerable; the whole inside of the gullet is white with them. I fix a little wooden prop between the two mandibles of the beak, to keep them open and enable me to see what happens.

I learn in this way that the hatching takes place in a couple of days. As soon as they are born, the young vermin, a swarming mass, leave the place where they are and disappear down the throat.

The beak of the bird invaded was closed at the start, as far as the natural contact of the mandibles allowed. There remained a narrow slit at the base, sufficient at most to admit the passage of a horse-hair. It was through this that the laying was performed. Lengthening her ovipositor like a telescope, the mother inserted the point of her implement, a point slightly hardened with a horny armour. The fineness of the probe equals the fineness of the aperture. But, if the beak were entirely closed, where would the eggs be laid then?

With a tied thread I keep the two mandibles in absolute contact; and I place a second Bluebottle in the presence of the Linnet, whom the colonists have already entered by the beak. This time the laying takes place on one of the eyes, between the lid and the eyeball. At the hatching, which again occurs a couple of days later, the grubs make their way into the fleshy depths of the socket. The eyes and the beak, therefore, form the two chief entrances into feathered game.

There are others; and these are the wounds. I cover the Linnet's head with a paper hood which will prevent invasion through the beak and eyes. I serve it, under the wire-gauze bell, to a third egg-layer. The bird has been struck by a shot in the breast, but the sore is not bleeding: no outer stain marks the injured spot. Moreover, I am careful to arrange the feathers, to smooth them with a hair-pencil, so that the bird looks quite smart and has every appearance of being untouched.

The Fly is soon there. She inspects the Linnet from end to end; with her front tarsi she fumbles at the breast and belly. It is a sort of auscultation by sense of touch. The insect becomes aware of what is under the feathers by the manner in which these react. If scent lends its assistance, it can only be very slightly, for the game is not yet high. The wound is soon found. No drop of blood is near it, for it is closed by a plug of down rammed into it by the shot. The Fly takes up her position without separating the feathers or uncovering the wound. She remains here for two hours without stirring, motionless, with her abdomen concealed beneath the plumage. My eager curiosity does not distract her from her business for a moment.

When she has finished, I take her place. There is nothing either on the skin or at the mouth of the wound. I have to withdraw the downy plug and dig to some depth before discovering the eggs. The ovipositor has therefore lengthened its extensible tube and pushed beyond the feather stopper driven in by the lead. The eggs are in one packet; they number about three hundred.

When the beak and eyes are rendered inaccessible, when the body, moreover, has no wounds, the laying still takes place, but this time in a hesitating and niggardly fashion. I pluck the bird completely, the better to watch what happens; also, I cover the head with a paper hood to close the usual means of access. For a long time, with jerky steps, the mother explores the body in every direction; she takes her stand by preference on the head, which she sounds by tapping on it with her front tarsi. She knows that the openings which she needs are there, under the paper; but she also knows how frail are her grubs, how powerless to pierce their way through the strange obstacle which stops her as well and interferes with the work of her ovipositor. The cowl inspires her with profound distrust. Despite the tempting bait of the veiled head, not an egg is laid on the wrapper, slight though it may be.

Weary of vain attempts to compass this obstacle, the Fly at last decides in favour of other points, but not on the breast, belly, or back, where the hide would seem too tough and the light too intrusive. She needs dark hiding-places, corners where the skin is very delicate. The spots chosen are the cavity of the axilla, corresponding with our arm-pit, and the crease where the thigh joins the belly. Eggs are laid in both places, but not many, showing that the groin and the axilla are adopted only reluctantly and for lack of a better spot.

With an unplucked bird, also hooded, the same experiment failed: the feathers prevent the Fly from slipping into those deep places. Let us add, in conclusion, that, on a skinned bird, or simply on a piece of butcher's meat, the laying is effected on any part whatever, provided that it be dark. The gloomiest corners are the favourite ones.

It follows from all this that, to lay her eggs, the Bluebottle picks out either naked wounds or else the mucous membranes of the mouth or eyes, which are not protected by a skin of any thickness. She also needs darkness.

The perfect efficiency of the paper bag, which prevents the inroads of the worms through the eye-sockets or the beak, suggests a similar experiment with the whole bird. It is a matter of wrapping the body in a sort of artificial skin which will be as discouraging to the Fly as the natural skin. Linnets, some with deep wounds, others almost intact, are placed one by one in paper envelopes similar to those in which the nursery-gardener keeps his seeds, envelopes just folded, without being stuck. The paper is quite ordinary and of middling thickness. Torn pieces of newspaper serve the purpose.

These sheaths with the corpses inside them are freely exposed to the air, on the table in my study, where they are visited, according to the time of day, in dense shade and in bright sunlight. Attracted by the effluvia from the dead meat, the Bluebottles haunt my laboratory, the windows of which are always open. I see them daily alighting on the envelopes and very busily exploring them, apprised of the contents by the gamy smell. Their incessant coming and going is a sign of intense cupidity; and yet none of them decides to lay on the bags. They do not even attempt to slide their ovipositor through the slits of the folds. The favourable season passes and not an egg is laid on the tempting wrappers. All the mothers abstain, judging the slender obstacle of the paper to be more than the vermin will be able to overcome.

This caution on the Fly's part does not at all surprise me: motherhood everywhere has great gleams of perspicacity. What does astonish me is the following result. The parcels containing the Linnets are left for a whole year uncovered on the table; they remain there for a second year and a third. I inspect the contents from time to time. The little birds are intact, with unrumpled feathers, free from smell, dry and light, like mummies. They have become not decomposed, but mummified.

I expected to see them putrefying, running into sanies, like corpses left to rot in the open air. On the contrary, the birds have dried and hardened, without undergoing any change. What did they want for their putrefaction? simply the intervention of the Fly. The maggot, therefore, is the primary cause of dissolution after death; it is, above all, the putrefactive chemist.

A conclusion not devoid of value may be drawn from my paper game-bags. In our markets, especially in those of the South, the game is hung unprotected from the hooks on the stalls. Larks strung up by the dozen with a wire through their nostrils, Thrushes, Plovers, Teal, Partridges, Snipe, in short, all the glories of the spit which the autumn migration brings us, remain for days and weeks at the mercy of the Flies. The buyer allows himself to be tempted by a goodly exterior; he makes his purchase and, back at home, just when the bird is being prepared for roasting, he discovers that the promised dainty is alive with worms. O horror! There is nothing for it but to throw the loathsome, verminous thing away.

The Bluebottle is the culprit here. Everybody knows it, and nobody thinks seriously of shaking off her tyranny: not the retailer, nor the wholesale dealer, nor the killer of the game. What is wanted to keep the maggots out? Hardly anything: to slip each bird into a paper sheath. If this precaution were taken at the start, before the Flies arrive, any game would be safe and could be left indefinitely to attain the degree of ripeness required by the epicure's palate.

Stuffed with olives and myrtleberries, the Corsican Blackbirds are exquisite eating. We sometimes receive them at Orange, layers of them, packed in baskets through which the air circulates freely and each contained in a paper wrapper. They are in a state of perfect preservation, complying with the most exacting demands of the kitchen. I congratulate the nameless shipper who conceived the bright idea of clothing his Blackbirds in paper. Will his example find imitators? I doubt it.

There is, of course, a serious objection to this method of preservation. In its paper shroud, the article is invisible; it is not enticing; it does not inform the passer-by of its nature and qualities. There is one resource left which would leave the bird uncovered: simply to case the head in a paper cap. The head being the part most menaced, because of the mucous membrane of the throat and eyes, it would be enough, as a rule, to protect the head, in order to keep off the Flies and thwart their attempts.

Let us continue to study the Bluebottle, while varying our means of information. A tin, about four inches deep, contains a piece of butcher's meat. The lid is not put in quite straight and leaves a narrow slit at one point of its circumference, allowing, at most, of the passage of a fine needle. When the bait begins to give off a gamy scent, the mothers come, singly or in numbers. They are attracted by the odour which, transmitted through a thin crevice, hardly reaches my nostrils.

They explore the metal receptacle for some time, seeking an entrance. Finding naught that enables them to reach the coveted morsel, they decide to lay their eggs on the tin, just beside the aperture. Sometimes, when the width of the passage allows of it, they insert the ovipositor into the tin and lay the eggs inside, on the very edge of the slit. Whether outside or in, the eggs are dabbed down in a fairly regular and absolutely white layer.

We have seen the Bluebottle refusing to lay her eggs on the paper bag, notwithstanding the carrion fumes of the Linnet enclosed; yet now, without hesitation, she lays them on a sheet of metal. Can the nature of the floor make any difference to her? I replace the tin lid by a paper cover stretched and pasted over the orifice. With the point of my knife I make a narrow slit in this new lid. That is quite enough: the parent accepts the paper.

What determined her, therefore, is not simply the smell, which can easily be perceived even through the uncut paper, but, above all, the crevice, which will provide an entrance for the vermin, hatched outside, near the narrow passage. The maggots' mother has her own logic, her prudent foresight. She knows how feeble her wee grubs will be, how powerless to cut their way through an obstacle of any resistance; and so, despite the temptation of the smell, she refrains from laying, so long as she finds no entrance through which the new-born worms can slip unaided.

I wanted to know whether the colour, the shininess, the degree of hardness and other qualities of the obstacle would influence the decision of a mother obliged to lay her eggs under exceptional conditions. With this object in view, I employed small jars, each baited with a bit of butcher's meat. The respective lids were made of different-coloured paper, of oil-skin, or of some of that tin-foil, with its gold or coppery sheen, which is used for sealing liqueur-bottles. On not one of these covers did the mothers stop, with any desire to deposit their eggs; but, from the moment that the knife had made the narrow slit, all the lids were, sooner or later, visited and all, sooner or later, received the white shower somewhere near the gash. The look of the obstacle, therefore, does not count; dull or brilliant, drab or coloured: these are details of no importance; the thing that matters is that there should be a passage to allow the grubs to enter.

Though hatched outside, at a distance from the coveted morsel, the new-born worms are well able to find their refectory. As they release themselves from the egg, without hesitation, so accurate is their scent, they slip beneath the edge of the ill-joined lid, or through the passage cut by the knife. Behold them entering upon their promised land, their reeking paradise.

Eager to arrive, do they drop from the top of the wall? Not they! Slowly creeping, they make their way down the side of the jar; they use their fore-part, ever in quest of information, as a crutch and grapnel in one. They reach the meat and at once instal themselves upon it.

Let us continue our investigation, varying the conditions. A large test-tube, measuring nine inches high, is baited at the bottom with a lump of butcher's meat. It is closed with wire-gauze, whose meshes, two millimetres wide (.078 inch.— Translator's Note.), do not permit of the Fly's passage. The Bluebottle comes to my apparatus, guided by scent rather than sight. She hastens to the test-tube, whose contents are veiled under an opaque cover, with the same alacrity as to the open tube. The invisible attracts her quite as much as the visible.

She stays awhile on the lattice of the mouth, inspects it attentively; but, whether because circumstances failed to serve me, or because the wire network inspired her with distrust, I never saw her dab her eggs upon it for certain. As her evidence was doubtful, I had recourse to the Flesh-fly (Sarcophaga carnaria).

This Fly is less finicking in her preparations, she has more faith in the strength of her worms, which are born ready-formed and vigorous, and easily shows me what I wish to see. She explores the trellis-work, chooses a mesh through which she inserts the tip of her abdomen, and, undisturbed by my presence, emits, one after the other, a certain number of grubs, about ten or so. True, her visits will be repeated, increasing the family at a rate of which I am ignorant.

The new-born worms, thanks to a slight viscidity, cling for a moment to the wire-gauze; they swarm, wriggle, release themselves and leap into the chasm. It is a nine-inch drop at least. When this is done, the mother makes off, knowing for a certainty that her offspring will shift for themselves. If they fall on the meat, well and good; if they fall elsewhere, they can reach the morsel by crawling.

This confidence in the unknown factor of the precipice, with no indication but that of smell, deserves fuller investigation. From what height will the Flesh-fly dare to let her children drop? I top the test-tube with another tube, the width of the neck of a claret-bottle. The mouth is closed either with wire-gauze or with a paper cover with a slight cut in it. Altogether, the apparatus measures twenty-five inches in height. No matter: the fall is not serious for the lithe backs of the young grubs; and, in a few days, the test-tube is filled with larvae, in which it is easy to recognize the Flesh-fly's family by the fringed coronet that opens and shuts at the maggot's stern like the petals of a little flower. I did not see the mother operating: I was not there at the time; but there is no doubt possible of her coming, nor of the great dive taken by the family: the contents of the test-tube furnish me with a duly authenticated certificate.

I admire the leap and, to obtain one better still, I replace the tube by another, so that the apparatus now stands forty-six inches high. The column is erected at a spot frequented by Flies, in a dim light. Its mouth, closed with a wire-gauze cover, reaches the level of various other appliances, test-tubes and jars, which are already stocked or awaiting their colony of vermin. When the position is well-known to the Flies, I remove the other tubes and leave the column, lest the visitors should turn aside to easier ground.

From time to time the Bluebottle and the Flesh-fly perch on the trellis-work, make a short investigation and then decamp. Throughout the summer season, for three whole months, the apparatus remains where it is, without result: never a worm. What is the reason? Does the stench of the meat not spread, coming from that depth? Certainly it spreads: it is unmistakable to my dulled nostrils and still more so to the nostrils of my children, whom I call to bear witness. Then why does the Flesh-fly, who but now was dropping her grubs from a goodly height, refuse to let them fall from the top of a column twice as high? Does she fear lest her worms should be bruised by an excessive drop? There is nothing about her to point to anxiety aroused by the length of the shaft. I never see her explore the tube or take its size. She stands on the trellised orifice; and there the matter ends. Can she be apprised of the depth of the chasm by the comparative faintness of the offensive odours that arise from it? Can the sense of smell measure the distance and judge whether it be acceptable or not? Perhaps.

The fact remains that, despite the attraction of the scent, the Flesh-fly does not expose her worms to disproportionate falls. Can she know beforehand that, when the chrysalids break, her winged family, knocking with a sudden flight against the sides of a tall chimney, will be unable to get out? This foresight would be in agreement with the rules which order maternal instinct according to future needs.

But, when the fall does not exceed a certain depth, the budding worms of the Flesh-fly are dropped without a qualm, as all our experiments show. This principle has a practical application which is not without its value in matters of domestic economy. It is as well that the wonders of entomology should sometimes give us a hint of commonplace utility.

The usual meat-safe is a sort of large cage with a top and bottom of wood and four wire-gauze sides. Hooks fixed into the top are used whereby to hang pieces which we wish to protect from the Flies. Often, so as to employ the space to the best advantage, these pieces are simply laid on the floor of the cage. With these arrangements, are we sure of warding off the Fly and her vermin?

Not at all. We may protect ourselves against the Bluebottle, who is not much inclined to lay her eggs at a distance from the meat; but there is still the Flesh-fly, who is more venturesome and goes more briskly to work and who will slip the grubs through a hole in the meshes and drop them inside the safe. Agile as they are and well able to crawl, the worms will easily reach anything on the floor; the only things secure from their attacks will be the pieces hanging from the ceiling. It is not in the nature of maggots to explore the heights, especially if this implies climbing down a string in addition.

People also use wire-gauze dish-covers. The trellised dome protects the contents even less than does the meat-safe. The Flesh-fly takes no heed of it. She can drop her worms through the meshes on the covered joint.

Then what are we to do? Nothing could be simpler. We need only wrap the birds which we wish to preserve—Thrushes, Partridges, Snipe and so on—in separate paper envelopes; and the same with our beef and mutton. This defensive armour alone, while leaving ample room for the air to circulate, makes any invasion by the worms impossible; even without a cover or a meat-safe: not that paper possesses any special preservative virtues, but solely because it forms an impenetrable barrier. The Bluebottle carefully refrains from laying her eggs upon it and the Flesh-fly from bringing forth her offspring, both of them knowing that their new-born young are incapable of piercing the obstacle.

Paper is equally successful in our strife against the Moths, those plagues of our furs and clothes. To keep away these wholesale ravagers, people generally use camphor, naphthalene, tobacco, bunches of lavender, and other strong-scented remedies. Without wishing to malign those preservatives, we are bound to admit that the means employed are none too effective. The smell does very little to prevent the havoc of the Moths.

I would therefore advise our housewives, instead of all this chemist's stuff, to use newspapers of a suitable shape and size. Take whatever you wish to protect—your furs, your flannel, or your clothes—and pack each article carefully in a newspaper, joining the edges with a double fold, well pinned. If this joining is properly done, the Moth will never get inside. Since my advice has been taken and this method employed in my household, the old damage has no longer been repeated.

To return to the Fly. A piece of meat is hidden in a jar under a layer of fine, dry sand, a finger's-breadth thick. The jar has a wide mouth and is left quite open. Let whoso come that will, attracted by the smell. The Bluebottles are not long in inspecting what I have prepared for them: they enter the jar, go out and come back again, inquiring into the invisible thing revealed by its fragrance. A diligent watch enables me to see them fussing about, exploring the sandy expanse, tapping it with their feet, sounding it with their proboscis. I leave the visitors undisturbed for a fortnight or three weeks. None of them lays any eggs.

This is a repetition of what the paper bag, with its dead bird, showed me. The Flies refuse to lay on the sand, apparently for the same reasons. The paper was considered an obstacle which the frail vermin would not be able to overcome. With sand, the case is worse. Its grittiness would hurt the new-born weaklings, its dryness would absorb the moisture indispensable to their movements. Later, when preparing for the metamorphosis, when their strength has come to them, the grubs will dig the earth quite well and be able to descend: but, at the start, that would be very dangerous for them. Knowing these difficulties, the mothers, however greatly tempted by the smell, abstain from breeding. As a matter of fact, after long waiting, fearing lest some packets of eggs may have escaped my attention, I inspect the contents of the jar from top to bottom. Meat and sand contain neither larvae nor pupae: the whole is absolutely deserted.

The layer of sand being only a finger's-breadth thick, this experiment requires certain precautions. The meat may expand a little, in going bad, and protrude in one or two places. However small the fleshy eyots that show above the surface, the Flies come to them and breed. Sometimes also the juices oozing from the putrid meat soak a small extent of the sandy floor. That is enough for the maggot's first establishment. These causes of failure are avoided with a layer of sand about an inch thick. Then the Bluebottle, the Flesh-fly, and other Flies whose grubs batten on dead bodies are kept at a proper distance.

In the hope of awakening us to a proper sense of our insignificance, pulpit orators sometimes make an unfair use of the grave and its worms. Let us put no faith in their doleful rhetoric. The chemistry of man's final dissolution is eloquent enough of our emptiness: there is no need to add imaginary horrors. The worm of the sepulchre is an invention of cantankerous minds, incapable of seeing things as they are. Covered by but a few inches of earth, the dead can sleep their quiet sleep: no Fly will ever come to take advantage of them.

At the surface of the soil, exposed to the air, the hideous invasion is possible; aye, it is the invariable rule. For the melting down and remoulding of matter, man is no better, corpse for corpse, than the lowest of the brutes. Then the Fly exercises her rights and deals with us as she does with any ordinary animal refuse. Nature treats us with magnificent indifference in her great regenerating factory: placed in her crucibles, animals and men, beggars and kings are 1 and all alike. There you have true equality, the only equality in this world of ours: equality in the presence of the maggot.

CHAPTER 8.

THE PINE-PROCESSIONARY.

Drover Dingdong's Sheep followed the Ram which Panurge had maliciously thrown overboard and leapt nimbly into the sea, one after the other, "for you know," says Rabelais, "it is the nature of the sheep always to follow the first, wheresoever it goes."

The Pine caterpillar is even more sheeplike, not from foolishness, but from necessity: where the first goes all the others go, in a regular string, with not an empty space between them.

They proceed in single file, in a continuous row, each touching with its head the rear of the one in front of it. The complex twists and turns described in his vagaries by the caterpillar leading the van are scrupulously described by all the others. No Greek theoria winding its way to the Eleusinian festivals was ever more orderly. Hence the name of Processionary given to the gnawer of the pine.

His character is complete when we add that he is a rope-dancer all his life long: he walks only on the tight-rope, a silken rail placed in position as he advances. The caterpillar who chances to be at the head of the procession dribbles his thread without ceasing and fixes it on the path which his fickle preferences cause him to take. The thread is so tiny that the eye, though armed with a magnifying-glass, suspects it rather than sees it.

But a second caterpillar steps on the slender foot-board and doubles it with his thread; a third trebles it; and all the others, however many there be, add the sticky spray from their spinnerets, so much so that, when the procession has marched by, there remains, as a record of its passing, a narrow white ribbon whose dazzling whiteness shimmers in the sun. Very much more sumptuous than ours, their system of road-making consists in upholstering with silk instead of macadamizing. We sprinkle our roads with broken stones and level them by the pressure of a heavy steam-roller; they lay over their paths a soft satin rail, a work of general interest to which each contributes his thread.

What is the use of all this luxury? Could they not, like other caterpillars, walk about without these costly preparations? I see two reasons for their mode of progression. It is night when the Processionaries sally forth to browse upon the pine-leaves. They leave their nest, situated at the top of a bough, in profound darkness; they go down the denuded pole till they come to the nearest branch that has not yet been gnawed, a branch which becomes lower and lower by degrees as the consumers finish stripping the upper storeys; they climb up this untouched branch and spread over the green needles.

When they have had their suppers and begin to feel the keen night air, the next thing is to return to the shelter of the house. Measured in a straight line, the distance is not great, hardly an arm's length; but it cannot be covered in this way on foot. The caterpillars have to climb down from one crossing to the next, from the needle to the twig, from the twig to the branch, from the branch to the bough and from the bough, by a no less angular path, to go back home. It is useless to rely upon sight as a guide on this long and erratic journey. The Processionary, it is true, has five ocular specks on either side of his head, but they are so infinitesimal, so difficult to make out through the magnifying-glass, that we cannot attribute to them any great power of vision. Besides, what good would those short-sighted lenses be in the absence of light, in black darkness?

It is equally useless to think of the sense of smell. Has the Processional any olfactory powers or has he not? I do not know. Without giving a positive answer to the question, I can at least declare that his sense of smell is exceedingly dull and in no way suited to help him find his way. This is proved, in my experiments, by a number of hungry caterpillars that, after a long fast, pass close beside a pine-branch without betraying any eagerness of showing a sign of stopping. It is the sense of touch that tells them where they are. So long as their lips do not chance to light upon the pasture-land, not one of them settles there, though he be ravenous. They do not hasten to food which they have scented from afar; they stop at a branch which they encounter on their way.

Apart from sight and smell, what remains to guide them in returning to the nest? The ribbon spun on the road. In the Cretan labyrinth, Theseus would have been lost but for the clue of thread with which Ariadne supplied him. The spreading maze of the pine-needles is, especially at night, as inextricable a labyrinth as that constructed for Minos. The Processionary finds his way through it, without the possibility of a mistake, by the aid of his bit of silk. At the time for going home, each easily recovers either his own thread or one or other of the neighbouring threads, spread fanwise by the diverging herd; one by one the scattered tribe line up on the common ribbon, which started from the nest; and the sated caravan finds its way back to the manor with absolute certainty.

Longer expeditions are made in the daytime, even in winter, if the weather be fine. Our caterpillars then come down from the tree, venture on the ground, march in procession for a distance of thirty yards or so. The object of these sallies is not to look for food, for the native pine-tree is far from being exhausted: the shorn branches hardly count amid the vast leafage. Moreover, the caterpillars observe complete abstinence till nightfall. The trippers have no other object than a constitutional, a pilgrimage to the outskirts to see what these are like, possibly an inspection of the locality where, later on, they mean to bury themselves in the sand for their metamorphosis.

It goes without saying that, in these greater evolutions, the guiding cord is not neglected. It is now more necessary than ever. All contribute to it from the produce of their spinnerets, as is the invariable rule whenever there is a progression. Not one takes a step forward without fixing to the path the thread from his lips.

If the series forming the procession be at all long, the ribbon is dilated sufficiently to make it easy to find; nevertheless, on the homeward journey, it is not picked up without some hesitation. For observe that the caterpillars when on the march never turn completely; to wheel round on their tight-rope is a method utterly unknown to them. In order therefore to regain the road already covered, they have to describe a zigzag whose windings and extent are determined by the leader's fancy. Hence come gropings and roamings which are sometimes prolonged to the point of causing the herd to spend the night out of doors. It is not a serious matter. They collect into a motionless cluster. To-morrow the search will start afresh and will sooner or later be successful. Oftener still the winding curve meets the guide-thread at the first attempt. As soon as the first caterpillar has the rail between his legs, all hesitation ceases; and the band makes for the nest with hurried steps.

The use of this silk-tapestried roadway is evident from a second point of view. To protect himself against the severity of the winter which he has to face when working, the Pine Caterpillar weaves himself a shelter in which he spends his bad hours, his days of enforced idleness. Alone, with none but the meagre resources of his silk-glands, he would find difficulty in protecting himself on the top of a branch buffeted by the winds. A substantial dwelling, proof against snow, gales and icy fogs, requires the cooperation of a large number. Out of the individual's piled-up atoms, the community obtains a spacious and durable establishment.

The enterprise takes a long time to complete. Every evening, when the weather permits, the building has to be strengthened and enlarged. It is indispensable, therefore, that the corporation of workers should not be dissolved while the stormy season continues and the insects are still in the caterpillar stage. But, without special arrangements, each nocturnal expedition at grazing-time would be a cause of separation. At that moment of appetite for food there is a return to individualism. The caterpillars become more or less scattered, settling singly on the branches around; each browses his pine-needle separately. How are they to find one another afterwards and become a community again?

The several threads left on the road make this easy. With that guide, every caterpillar, however far he may be, comes back to his companions without ever missing the way. They come hurrying from a host of twigs, from here, from there, from above, from below; and soon the scattered legion reforms into a group. The silk thread is something more than a road-making expedient: it is the social bond, the system that keeps the members of the brotherhood indissolubly united.

At the head of every procession, long or short, goes a first caterpillar whom I will call the leader of the march or file, though the word leader, which I use for the want of a better, is a little out of place here. Nothing, in fact, distinguishes this caterpillar from the others: it just depends upon the order in which they happen to line up; and mere chance brings him to the front. Among the Processionaries, every captain is an officer of fortune. The actual leader leads; presently he will be a subaltern, if the line should break up in consequence of some accident and be formed anew in a different order.

His temporary functions give him an attitude of his own. While the others follow passively in a close file, he, the captain, tosses himself about and with an abrupt movement flings the front of his body hither and thither. As he marches ahead he seems to be seeking his way. Does he in point of fact explore the country? Does he choose the most practicable places? Or are his hesitations merely the result of the absence of a guiding thread on ground that has not yet been covered? His subordinates follow very placidly, reassured by the cord which they hold between their legs; he, deprived of that support, is uneasy.

Why cannot I read what passes under his black, shiny skull, so like a drop of tar to look at? To judge by actions, there is here a modicum of discernment which is able, after experimenting, to recognize excessive roughnesses, over-slippery surfaces, dusty places that offer no resistance and, above all, the threads left by other excursionists. This is all or nearly all that my long acquaintance with the Processionaries has taught me as to their mentality. Poor brains, indeed; poor creatures, whose commonwealth has its safety hanging upon a thread!

The processions vary greatly in length. The finest that I have seen manoeuvring on the ground measured twelve or thirteen yards and numbered about three hundred caterpillars, drawn up with absolute precision in a wavy line. But, if there were only two in a row the order would still be perfect: the second touches and follows the first.

By February I have processions of all lengths in the greenhouse. What tricks can I play upon them? I see only two: to do away with the leader; and to cut the thread.

The suppression of the leader of the file produces nothing striking. If the thing is done without creating a disturbance, the procession does not alter its ways at all. The second caterpillar, promoted to captain, knows the duties of his rank offhand: he selects and leads, or rather he hesitates and gropes.

The breaking of the silk ribbon is not very important either. I remove a caterpillar from the middle of the file. With my scissors, so as not to cause a commotion in the ranks, I cut the piece of ribbon on which he stood and clear away every thread of it. As a result of this breach, the procession acquires two marching leaders, each independent of the other. It may be that the one in the rear joins the file ahead of him, from which he is separated by but a slender interval; in that case, things return to their original condition. More frequently, the two parts do not become reunited. In that case, we have two distinct processions, each of which wanders where it pleases and diverges from the other. Nevertheless, both will be able to return to the nest by discovering sooner or later, in the course of their peregrinations, the ribbon on the other side of the break.

These two experiments are only moderately interesting. I have thought out another, one more fertile in possibilities. I propose to make the caterpillars describe a close circuit, after the ribbons running from it and liable to bring about a change of direction have been destroyed. The locomotive engine pursues its invariable course so long as it is not shunted on to a branch-line. If the Processionaries find the silken rail always clear in front of them, with no switches anywhere, will they continue on the same track, will they persist in following a road that never comes to an end? What we have to do is to produce this circuit, which is unknown under ordinary conditions, by artificial means.

The first idea that suggests itself is to seize with the forceps the silk ribbon at the back of the train, to bend it without shaking it and to bring the end of it ahead of the file. If the caterpillar marching in the van steps upon it, the thing is done: the others will follow him faithfully. The operation is very simple in theory but most difficult in practice and produces no useful results. The ribbon, which is extremely slight, breaks under the weight of the grains of sand that stick to it and are lifted with it. If it does not break, the caterpillars at the back, however delicately we may go to work, feel a disturbance which makes them curl up or even let go.

There is a yet greater difficulty: the leader refuses the ribbon laid before him; the cut end makes him distrustful. Failing to see the regular, uninterrupted road, he slants off to the right or left, he escapes at a tangent. If I try to interfere and to bring him back to the path of my choosing, he persists in his refusal, shrivels up, does not budge, and soon the whole procession is in confusion. We will not insist: the method is a poor one, very wasteful of effort for at best a problematical success.

We ought to interfere as little as possible and obtain a natural closed circuit. Can it be done? Yes. It lies in our power, without the least meddling, to see a procession march along a perfect circular track. I owe this result, which is eminently deserving of our attention, to pure chance.

On the shelf with the layer of sand in which the nests are planted stand some big palm-vases measuring nearly a yard and a half in circumference at the top. The caterpillars often scale the sides and climb up to the moulding which forms a cornice around the opening. This place suits them for their processions, perhaps because of the absolute firmness of the surface, where there is no fear of landslides, as on the loose, sandy soil below; and also, perhaps, because of the horizontal position, which is favourable to repose after the fatigue of the ascent. It provides me with a circular track all ready-made. I have nothing to do but wait for an occasion propitious to my plans. This occasion is not long in coming.

On the 30th of January, 1896, a little before twelve o'clock in the day, I discover a numerous troop making their way up and gradually reaching the popular cornice. Slowly, in single file, the caterpillars climb the great vase, mount the ledge and advance in regular procession, while others are constantly arriving and continuing the series. I wait for the string to close up, that is to say, for the leader, who keeps following the circular moulding, to return to the point from which he started. My object is achieved in a quarter of an hour. The closed circuit is realized magnificently, in something very nearly approaching a circle.

The next thing is to get rid of the rest of the ascending column, which would disturb the fine order of the procession by an excess of newcomers; it is also important that we should do away with all the silken paths, both new and old, that can put the cornice into communication with the ground. With a thick hair-pencil I sweep away the surplus climbers; with a big brush, one that leaves no smell behind it—for this might afterwards prove confusing—I carefully rub down the vase and get rid of every thread which the caterpillars have laid on the march. When these preparations are finished, a curious sight awaits us.

In the interrupted circular procession there is no longer a leader. Each caterpillar is preceded by another on whose heels he follows guided by the silk track, the work of the whole party; he again has a companion close behind him, following him in the same orderly way. And this is repeated without variation throughout the length of the chain. None commands, or rather none modifies the trail according to his fancy; all obey, trusting in the guide who ought normally to lead the march and who in reality has been abolished by my trickery.

From the first circuit of the edge of the tub the rail of silk has been laid in position and is soon turned into a narrow ribbon by the procession, which never ceases dribbling its thread as it goes. The rail is simply doubled and has no branches anywhere, for my brush has destroyed them all. What will the caterpillars do on this deceptive, closed path? Will they walk endlessly round and round until their strength gives out entirely?

The old schoolmen were fond of quoting Buridan's Ass, that famous Donkey who, when placed between two bundles of hay, starved to death because he was unable to decide in favour of either by breaking the equilibrium between two equal but opposite attractions. They slandered the worthy animal. The Ass, who is no more foolish than any one else, would reply to the logical snare by feasting off both bundles. Will my caterpillars show a little of his mother wit? Will they, after many attempts, be able to break the equilibrium of their closed circuit, which keeps them on a road without a turning? Will they make up their minds to swerve to this side or that, which is the only method of reaching their bundle of hay, the green branch yonder, quite near, not two feet off?

I thought that they would and I was wrong. I said to myself:

"The procession will go on turning for some time, for an hour, two hours, perhaps; then the caterpillars will perceive their mistake. They will abandon the deceptive road and make their descent somewhere or other."

That they should remain up there, hard pressed by hunger and the lack of cover, when nothing prevented them from going away, seemed to me inconceivable imbecility. Facts, however, forced me to accept the incredible. Let us describe them in detail.

The circular procession begins, as I have said, on the 30th of January, about midday, in splendid weather. The caterpillars march at an even pace, each touching the stern of the one in front of him. The unbroken chain eliminates the leader with his changes of direction; and all follow mechanically, as faithful to their circle as are the hands of a watch. The headless file has no liberty left, no will; it has become mere clockwork. And this continues for hours and hours. My success goes far beyond my wildest suspicions. I stand amazed at it, or rather I am stupefied.

Meanwhile, the multiplied circuits change the original rail into a superb ribbon a twelfth of an inch broad. I can easily see it glittering on the red ground of the pot. The day is drawing to a close and no alteration has yet taken place in the position of the trail. A striking proof confirms this.

The trajectory is not a plane curve, but one which, at a certain point, deviates and goes down a little way to the lower surface of the cornice, returning to the top some eight inches farther. I marked these two points of deviation in pencil on the vase at the outset. Well, all that afternoon and, more conclusive still, on the following days, right to the end of this mad dance, I see the string of caterpillars dip under the ledge at the first point and come to the top again at the second. Once the first thread is laid, the road to be pursued is permanently established.

If the road does not vary, the speed does. I measure nine centimetres (3 1/2 inches.—Translator's Note.) a minute as the average distance covered. But there are more or less lengthy halts; the pace slackens at times, especially when the temperature falls. At ten o'clock in the evening the walk is little more than a lazy swaying of the body. I foresee an early halt, in consequence of the cold, of fatigue and doubtless also of hunger.

Grazing-time has arrived. The caterpillars have come crowding from all the nests in the greenhouse to browse upon the pine-branches planted by myself beside the silken purses. Those in the garden do the same, for the temperature is mild. The others, lined up along the earthenware cornice, would gladly take part in the feast;

they are bound to have an appetite after a ten hours' walk. The branch stands green and tempting not a hand's-breadth away. To reach it they need but go down; and the poor wretches, foolish slaves of their ribbon that they are, cannot make up their minds to do so. I leave the famished ones at half-past ten, persuaded that they will take counsel with their pillow and that on the morrow things will have resumed their ordinary course.

I was wrong. I was expecting too much of them when I accorded them that faint gleam of intelligence which the tribulations of a distressful stomach ought, one would think, to have aroused. I visit them at dawn. They are lined up as on the day before, but motionless. When the air grows a little warmer, they shake off their torpor, revive and start walking again. The circular procession begins anew, like that which I have already seen. There is nothing more and nothing less to be noted in their machine-like obstinacy.

This time it is a bitter night. A cold snap has supervened, was indeed foretold in the evening by the garden caterpillars, who refused to come out despite appearances which to my duller senses seemed to promise a continuation of the fine weather. At daybreak the rosemary-walks are all asparkle with rime and for the second time this year there is a sharp frost. The large pond in the garden is frozen over. What can the caterpillars in the conservatory be doing? Let us go and see.

All are ensconced in their nests, except the stubborn processionists on the edge of the vase, who, deprived of shelter as they are, seem to have spent a very bad night. I find them clustered in two heaps, without any attempt at order. They have suffered less from the cold, thus huddled together.

'Tis an ill wind that blows nobody any good. The severity of the night has caused the ring to break into two segments which will, perhaps, afford a chance of safety. Each group, as it survives and resumes its walk, will presently be headed by a leader who, not being obliged to follow a caterpillar in front of him, will possess some liberty of movement and perhaps be able to make the procession swerve to one side. Remember that, in the ordinary processions, the caterpillar walking ahead acts as a scout. While the others, if nothing occurs to create excitement, keep to their ranks, he attends to his duties as a leader and is continually turning his head to this side and that, investigating, seeking, groping, making his choice. And things happen as he decides: the band follows him faithfully. Remember also that, even on a road which has already been travelled and beribboned, the guiding caterpillar continues to explore.

There is reason to believe that the Processionaries who have lost their way on the ledge will find a chance of safety here. Let us watch them. On recovering from their torpor, the two groups line up by degrees into two distinct files. There are therefore two leaders, free to go where they please, independent of each other. Will they succeed in leaving the enchanted circle? At the sight of their large black heads swaying anxiously from side to side, I am inclined to think so for a moment. But I am soon undeceived. As the ranks fill out, the two sections of the chain meet and the circle is reconstituted. The momentary leaders once more become simple subordinates; and again the caterpillars march round and round all day.

For the second time in succession, the night, which is very calm and magnificently starry, brings a hard frost. In the morning the Processionaries on the tub, the only ones who have camped unsheltered, are gathered into a heap which largely overflows both sides of the fatal ribbon. I am present at the awakening of the numbed ones. The first to take the road is, as luck will have it, outside the track. Hesitatingly he ventures into unknown ground. He reaches the top of the rim and descends upon the other side on the earth in the vase. He is followed by six others, no more. Perhaps the rest of the troop, who have not fully recovered from their nocturnal torpor, are too lazy to bestir themselves.

The result of this brief delay is a return to the old track. The caterpillars embark on the silken trail and the circular march is resumed, this time in the form of a ring with a gap in it. There is no attempt, however, to strike a new course on the part of the guide whom this gap has placed at the head. A chance of stepping outside the magic circle has presented itself at last; and he does not know how to avail himself of it.

As for the caterpillars who have made their way to the inside of the vase, their lot is hardly improved. They climb to the top of the palm, starving and seeking for food. Finding nothing to eat that suits them, they retrace their steps by following the thread which they have left on the way, climb the ledge of the pot, strike the procession again and, without further anxiety, slip back into the ranks. Once more the ring is complete, once more the circle turns and turns.

Then when will the deliverance come? There is a legend that tells of poor souls dragged along in an endless round until the hellish charm is broken by a drop of holy water. What drop will good fortune sprinkle on my Processionaries to dissolve their circle and bring them back to the nest? I see only two means of conjuring the spell and obtaining a release from the circuit. These two means are two painful ordeals. A strange linking of cause and effect: from sorrow and wretchedness good is to come.

And, first, shriveling as the result of cold, the caterpillars gather together without any order, heap themselves some on the path, some, more numerous these, outside it. Among the latter there may be, sooner or later, some revolutionary who, scorning the beaten track, will trace out a new road and lead the troop back home. We have just seen an instance of it. Seven penetrated to the interior of the vase and climbed the palm. True, it was an attempt with no result but still an attempt. For complete success, all that need be done would have been to take the opposite slope. An even chance is a great thing. Another time we shall be more successful.

In the second place, the exhaustion due to fatigue and hunger. A lame one stops, unable to go farther. In front of the defaulter the procession still continues to wend its way for a short time. The ranks close up and an empty space appears. On coming to himself and resuming the march, the caterpillar who has caused the breach becomes a leader, having nothing before him. The least desire for emancipation is all that he wants to make him launch the band into a new path which perhaps will be the saving path.

In short, when the Processionaries' train is in difficulties, what it needs, unlike ours, is to run off the rails. The side-tracking is left to the caprice of a leader who alone is capable of turning to the right or left; and this leader is absolutely non-existent so long as the ring remains unbroken. Lastly, the breaking of the circle, the one stroke of luck, is the result of a chaotic halt, caused principally by excess of fatigue or cold.

The liberating accident, especially that of fatigue, occurs fairly often. In the course of the same day, the moving circumference is cut up several times into two or three sections; but continuity soon returns and no change takes place. Things go on just the same. The bold innovator who is to save the situation has not yet had his inspiration.

There is nothing new on the fourth day, after an icy night like the previous one; nothing to tell except the following detail. Yesterday I did not remove the trace left by the few caterpillars who made their way to the inside of the vase. This trace, together with a junction connecting it with the circular road, is discovered in the course of the morning. Half the troop takes advantage of it to visit the earth in the pot and climb the palm; the other half remains on the ledge and continues to walk along the old rail. In the afternoon the band of emigrants rejoins the others, the circuit is completed and things return to their original condition.

We come to the fifth day. The night frost becomes more intense, without however as yet reaching the greenhouse. It is followed by bright sunshine in a calm and limpid sky. As soon as the sun's rays have warmed the panes a little, the caterpil-lars, lying in heaps, wake up and resume their evolutions on the ledge of the vase. This time the fine order of the beginning is disturbed and a certain disorder becomes manifest, apparently an omen of deliverance near at hand. The scouting-path inside the vase, which was upholstered in silk yesterday and the day before, is to-day followed to its origin on the rim by a part of the band and is then desert-ed after a short loop. The other caterpillars follow the usual ribbon. The result of this bifurcation is two almost equal files, walking along the ledge in the same direction, at a short distance from each other, sometimes meeting, separating far-ther on, in every case with some lack of order.

Weariness increases the confusion. The crippled, who refuse to go on, are many. Breaches increase; files are split up into sections each of which has its leader, who pokes the front of his body this way and that to explore the ground. Everything seems to point to the disintegration which will bring safety. My hopes are once more disappointed. Before the night the single file is reconstituted and the invin-cible gyration resumed.

Heat comes, just as suddenly as the cold did. To-day, the 4th of February, is a beautiful, mild day. The greenhouse is full of life. Numerous festoons of caterpil-lars, issuing from the nests, meander along the sand on the shelf. Above them, at every moment, the ring on the ledge of the vase breaks up and comes together again. For the first time I see daring leaders who, drunk with heat, standing only on their hinder prolegs at the extreme edge of the earthenware rim, fling them-selves forward into space, twisting about, sounding the depths. The endeavour is frequently repeated, while the whole troop stops. The caterpillars' heads give sud-den jerks, their bodies wriggle.

One of the pioneers decides to take the plunge. He slips under the ledge. Four fol-low him. The others, still confiding in the perfidious silken path, dare not copy him and continue to go along the old road.

The short string detached from the general chain gropes about a great deal, hesi-tates long on the side of the vase; it goes half-way down, then climbs up again slantwise, rejoins and takes its place in the procession. This time the attempt has failed, though at the foot of the vase, not nine inches away, there lay a bunch of pine-needles which I had placed there with the object of enticing the hungry ones. Smell and sight told them nothing. Near as they were to the goal, they went up again.

No matter, the endeavour has its uses. Threads were laid on the way and will serve as a lure to further enterprise. The road of deliverance has its first landmarks. And, two days later, on the eighth day of the experiment, the caterpillars—now singly, anon in small groups, then again in strings of some length—come down from the ledge by following the staked-out path. At sunset the last of the laggards is back in the nest.

Now for a little arithmetic. For seven times twenty-four hours the caterpillars have remained on the ledge of the vase. To make an ample allowance for stops due to the weariness of this one or that and above all for the rest taken during the colder hours of the night, we will deduct one-half of the time. This leaves eighty-four hours' walking. The average pace is nine centimetres a minute. (3 1/2 inches.—Translator's Note.) The aggregate distance covered, therefore, is 453 metres, a good deal more than a quarter of a mile, which is a great walk for these little crawlers. The circumference of the vase, the perimeter of the track, is exactly 1 metre 35. (4 feet 5 inches.—Translator's Note.) Therefore the circle covered, always in the same direction and always without result, was described three hundred and thirty-five times.

These figures surprise me, though I am already familiar with the abysmal stupidity of insects as a class whenever the least accident occurs. I feel inclined to ask myself whether the Processionaries were not kept up there so long by the difficulties and dangers of the descent rather than by the lack of any gleam of intelligence in their benighted minds. The facts, however, reply that the descent is as easy as the ascent.

The caterpillar has a very supple back, well adapted for twisting round projections or slipping underneath. He can walk with the same ease vertically or horizontally, with his back down or up. Besides, he never moves forward until he has fixed his thread to the ground. With this support to his feet, he has no falls to fear, no matter what his position.

I had a proof of this before my eyes during a whole week. As I have already said, the track, instead of keeping on one level, bends twice, dips at a certain point under the ledge of the vase and reappears at the top a little farther on. At one part of the circuit, therefore, the procession walks on the lower surface of the rim; and this inverted position implies so little discomfort or danger that it is renewed at each turn for all the caterpillars from first to last.

It is out of the question then to suggest the dread of a false step on the edge of the rim which is so nimbly turned at each point of inflexion. The caterpillars in distress, starved, shelterless, chilled with cold at night, cling obstinately to the silk ribbon covered hundreds of times, because they lack the rudimentary glimmers of reason which would advise them to abandon it.

Experience and reflection are not in their province. The ordeal of a five hundred yards' march and three to four hundred turns teach them nothing; and it takes casual circumstances to bring them back to the nest. They would perish on their insidious ribbon if the disorder of the nocturnal encampments and the halts due to fatigue did not cast a few threads outside the circular path. Some three or four move along these trails, laid without an object, stray a little way and, thanks to their wanderings, prepare the descent, which is at last accomplished in short strings favoured by chance.

The school most highly honoured to-day is very anxious to find the origin of reason in the dregs of the animal kingdom. Let me call its attention to the Pine Processionary.

CHAPTER 9.

THE SPIDERS.

THE NARBONNE LYCOSA, OR BLACK-BELLIED TARANTULA.

THE BURROW.

Michelet has told us how, as a printer's apprentice in a cellar, he established amicable relations with a Spider. (Jules Michelet (1798-1874), author of "L'Oiseau" and "L'Insecte," in addition to the historical works for which he is chiefly known. As a lad, he helped his father, a printer by trade, in setting type.—Translator's Note.) At a certain hour of the day, a ray of sunlight would glint through the window of the gloomy workshop and light up the little compositor's case. Then his eight-legged neighbour would come down from her web and on the edge of the case take her share of the sunshine. The boy did not interfere with her; he welcomed the trusting visitor as a friend and as a pleasant diversion from the long monotony. When we lack the society of our fellow-men, we take refuge in that of animals, without always losing by the change.

I do not, thank God, suffer from the melancholy of a cellar: my solitude is gay with light and verdure; I attend, whenever I please, the fields' high festival, the Thrushes' concert, the Crickets' symphony; and yet my friendly commerce with the Spider is marked by an even greater devotion than the young type-setter's. I admit her to the intimacy of my study, I make room for her among my books, I set her in the sun on my window-ledge, I visit her assiduously at her home, in the country. The object of our relations is not to create a means of escape from the petty worries of life, pin-pricks whereof I have my share like other men, a very large share, indeed; I propose to submit to the Spider a host of questions whereto, at times, she condescends to reply.

To what fair problems does not the habit of frequenting her give rise! To set them forth worthily, the marvellous art which the little printer was to acquire were not too much. One needs the pen of a Michelet; and I have but a rough, blunt pencil. Let us try, nevertheless: even when poorly clad, truth is still beautiful.

The most robust Spider in my district is the Narbonne Lycosa, or Black-bellied Tarantula, clad in black velvet on the lower surface, especially under the belly, with brown chevrons on the abdomen and grey and white rings around the legs. Her favourite home is the dry, pebbly ground, covered with sun-scorched thyme. In my harmas laboratory there are quite twenty of this Spider's burrows. Rarely do I pass by one of these haunts without giving a glance down the pit where gleam, like diamonds, the four great eyes, the four telescopes, of the hermit. The four others, which are much smaller, are not visible at that depth.

Would I have greater riches, I have but to walk a hundred yards from my house, on the neighbouring plateau, once a shady forest, to-day a dreary solitude where the Cricket browses and the Wheat-ear flits from stone to stone. The love of lucre has laid waste the land. Because wine paid handsomely, they pulled up the forest to plant the vine. Then came the Phylloxera, the vine-stocks perished and the once green table-land is now no more than a desolate stretch where a few tufts of hardy grasses sprout among the pebbles. This waste-land is the Lycosa's paradise: in an hour's time, if need were, I should discover a hundred burrows within a limited range.

These dwellings are pits about a foot deep, perpendicular at first and then bent elbow-wise. The average diameter is an inch. On the edge of the hole stands a kerb, formed of straw, bits and scraps of all sorts and even small pebbles, the size of a hazel-nut. The whole is kept in place and cemented with silk. Often, the Spider confines herself to drawing together the dry blades of the nearest grass, which she ties down with the straps from her spinnerets, without removing the blades from the stems; often, also, she rejects this scaffolding in favour of a masonry constructed of small stones. The nature of the kerb is decided by the nature of the materials within the Lycosa's reach, in the close neighbourhood of the building-yard. There is no selection: everything meets with approval, provided that it be near at hand.

The direction is perpendicular, in so far as obstacles, frequent in a soil of this kind, permit. A bit of gravel can be extracted and hoisted outside; but a flint is an immovable boulder which the Spider avoids by giving a bend to her gallery. If more such are met with, the residence becomes a winding cave, with stone vaults, with lobbies communicating by means of sharp passages.

This lack of plan has no attendant drawbacks, so well does the owner, from long habit, know every corner and storey of her mansion. If any interesting buzz occur overhead, the Lycosa climbs up from her rugged manor with the same speed as from a vertical shaft. Perhaps she even finds the windings and turnings an advantage, when she has to drag into her den a prey that happens to defend itself.

As a rule, the end of the burrow widens into a side-chamber, a lounge or resting-place where the Spider meditates at length and is content to lead a life of quiet when her belly is full.

When she reaches maturity and is once settled, the Lycosa becomes eminently domesticated. I have been living in close communion with her for the last three years. I have installed her in large earthen pans on the window-sills of my study and I have her daily under my eyes. Well, it is very rarely that I happen on her outside, a few inches from her hole, back to which she bolts at the least alarm.

We may take it then that, when not in captivity, the Lycosa does not go far afield to gather the wherewithal to build her parapet and that she makes shift with what she finds upon her threshold. In these conditions, the building-stones are soon exhausted and the masonry ceases for lack of materials.

The wish came over me to see what dimensions the circular edifice would assume, if the Spider were given an unlimited supply. With captives to whom I myself act as purveyor the thing is easy enough. Were it only with a view to helping whoso may one day care to continue these relations with the big Spider of the waste-lands, let me describe how my subjects are housed.

A good-sized earthenware pan, some nine inches deep, is filled with a red, clayey earth, rich in pebbles, similar, in short, to that of the places haunted by the Lycosa. Properly moistened into a paste, the artificial soil is heaped, layer by layer, around a central reed, of a bore equal to that of the animal's natural burrow. When the receptacle is filled to the top, I withdraw the reed, which leaves a yawning, perpendicular shaft. I thus obtain the abode which shall replace that of the fields.

To find the hermit to inhabit it is merely the matter of a walk in the neighbourhood. When removed from her own dwelling, which is turned topsy-turvy by my trowel, and placed in possession of the den produced by my art, the Lycosa at once disappears into that den. She does not come out again, seeks nothing better elsewhere. A large wire-gauze cover rests on the soil in the pan and prevents escape.

In any case, the watch, in this respect, makes no demand upon my diligence. The prisoner is satisfied with her new abode and manifests no regret for her natural burrow. There is no attempt at flight on her part. Let me not omit to add that each pan must receive not more than one inhabitant. The Lycosa is very intolerant. To her a neighbour is fair game, to be eaten without scruple when one has might on one's side. Time was when, unaware of this fierce intolerance, which is more savage still at breeding time, I saw hideous orgies perpetrated in my overstocked cages. I shall have occasion to describe those tragedies later.

Let us meanwhile consider the isolated Lycosae. They do not touch up the dwelling which I have moulded for them with a bit of reed; at most, now and again, perhaps with the object of forming a lounge or bedroom at the bottom, they fling out a few loads of rubbish. But all, little by little, build the kerb that is to edge the mouth.

I have given them plenty of first-rate materials, far superior to those which they use when left to their own resources. These consist, first, for the foundations, of little smooth stones, some of which are as large as an almond. With this road-metal are mingled short strips of raphia, or palm-fibre, flexible ribbons, easily bent. These stand for the Spider's usual basket-work, consisting of slender stalks and dry blades of grass. Lastly, by way of an unprecedented treasure, never yet employed by a Lycosa, I place at my captives' disposal some thick threads of wool, cut into inch lengths.

As I wish, at the same time, to find out whether my animals, with the magnificent lenses of their eyes, are able to distinguish colours and prefer one colour to another, I mix up bits of wool of different hues: there are red, green, white, and yellow pieces. If the Spider have any preference, she can choose where she pleases.

The Lycosa always works at night, a regrettable circumstance, which does not allow me to follow the worker's methods. I see the result; and that is all. Were I to visit the building-yard by the light of a lantern, I should be no wiser. The Spider, who is very shy, would at once dive into her lair; and I should have lost my sleep for nothing. Furthermore, she is not a very diligent labourer; she likes to take her time. Two or three bits of wool or raphia placed in position represent a whole night's work. And to this slowness we must add long spells of utter idleness.

Two months pass; and the result of my liberality surpasses my expectations. Possessing more windfalls than they know what to do with, all picked up in their immediate neighbourhood, my Lycosae have built themselves donjon-keeps the like of which their race has not yet known. Around the orifice, on a slightly sloping bank, small, flat, smooth stones have been laid to form a broken, flagged pavement. The larger stones, which are Cyclopean blocks compared with the size of the animal that has shifted them, are employed as abundantly as the others.

On this rockwork stands the donjon. It is an interlacing of raphia and bits of wool, picked up at random, without distinction of shade. Red and white, green and yellow are mixed without any attempt at order. The Lycosa is indifferent to the joys of colour.

The ultimate result is a sort of muff, a couple of inches high. Bands of silk, supplied by the spinnerets, unite the pieces, so that the whole resembles a coarse fabric. Without being absolutely faultless, for there are always awkward pieces on the outside, which the worker could not handle, the gaudy building is not devoid of merit. The bird lining its nest would do no better. Whoso sees the curious, many-coloured productions in my pans takes them for an outcome of my industry, contrived with a view to some experimental mischief; and his surprise is great when I confess who the real author is. No one would ever believe the Spider capable of constructing such a monument.

It goes without saying that, in a state of liberty, on our barren waste-lands, the Lycosa does not indulge in such sumptuous architecture. I have given the reason: she is too great a stay-at-home to go in search of materials and she makes use of the limited resources which she finds around her. Bits of earth, small chips of stone, a few twigs, a few withered grasses: that is all, or nearly all. Wherefore the work is generally quite modest and reduced to a parapet that hardly attracts attention.

My captives teach us that, when materials are plentiful, especially textile materials that remove all fears of landslip, the Lycosa delights in tall turrets. She understands the art of donjon-building and puts it into practice as often as she possesses the means.

What is the purpose of this turret? My pans will tell us that. An enthusiastic votary of the chase, so long as she is not permanently fixed, the Lycosa, once she has set up house, prefers to lie in ambush and wait for the quarry. Every day, when the heat is greatest, I see my captives come up slowly from under ground and lean upon the battlements of their woolly castle-keep. They are then really magnificent in their stately gravity. With their swelling belly contained within the aperture, their head outside, their glassy eyes staring, their legs gathered for a spring, for hours and hours they wait, motionless, bathing voluptuously in the sun.

Should a tit-bit to her liking happen to pass, forthwith the watcher darts from her tall tower, swift as an arrow from the bow. With a dagger-thrust in the neck, she stabs the jugular of the Locust, Dragon-fly or other prey whereof I am the purveyor; and she as quickly scales the donjon and retires with her capture. The performance is a wonderful exhibition of skill and speed.

Very seldom is a quarry missed, provided that it pass at a convenient distance, within the range of the huntress' bound. But, if the prey be at some distance, for instance on the wire of the cage, the Lycosa takes no notice of it. Scorning to go in pursuit, she allows it to roam at will. She never strikes except when sure of her stroke. She achieves this by means of her tower. Hiding behind the wall, she sees the stranger advancing, keeps her eyes on him and suddenly pounces when he comes within reach. These abrupt tactics make the thing a certainty. Though he were winged and swift of flight, the unwary one who approaches the ambush is lost.

This presumes, it is true, an exemplary patience on the Lycosa's part; for the burrow has naught that can serve to entice victims. At best, the ledge provided by the turret may, at rare intervals, tempt some weary wayfarer to use it as a resting-place. But, if the quarry do not come to-day, it is sure to come to-morrow, the next day, or later, for the Locusts hop innumerable in the waste-land, nor are they always able to regulate their leaps. Some day or other, chance is bound to bring one of them within the purlieus of the burrow. This is the moment to spring upon the pilgrim from the ramparts. Until then, we maintain a stoical vigilance. We shall dine when we can; but we shall end by dining.

The Lycosa, therefore, well aware of these lingering eventualities, waits and is not unduly distressed by a prolonged abstinence. She has an accommodating stomach, which is satisfied to be gorged to-day and to remain empty afterwards for goodness knows how long. I have sometimes neglected my catering duties for weeks at a time; and my boarders have been none the worse for it. After a more or less protracted fast, they do not pine away, but are smitten with a wolf-like hunger. All these ravenous eaters are alike: they guzzle to excess to-day, in anticipation of to-morrow's dearth.

THE LAYING.

Chance, a poor stand-by, sometimes contrives very well. At the beginning of the month of August, the children call me to the far side of the enclosure, rejoicing in a find which they have made under the rosemary-bushes. It is a magnificent Lycosa, with an enormous belly, the sign of an impending delivery.

Early one morning, ten days later, I find her preparing for her confinement. A silk network is first spun on the ground, covering an extent about equal to the palm of one's hand. It is coarse and shapeless, but firmly fixed. This is the floor on which the Spider means to operate.

On this foundation, which acts as a protection from the sand, the Lycosa fashions a round mat, the size of a two-franc piece and made of superb white silk. With a gentle, uniform movement, which might be regulated by the wheels of a delicate piece of clockwork, the tip of the abdomen rises and falls, each time touching the supporting base a little farther away, until the extreme scope of the mechanism is attained.

Then, without the Spider's moving her position, the oscillation is resumed in the opposite direction. By means of this alternate motion, interspersed with numerous contacts, a segment of the sheet is obtained, of a very accurate texture. When this is done, the Spider moves a little along a circular line and the loom works in the same manner on another segment.

The silk disk, a sort of hardy concave paten, now no longer receives anything from the spinnerets in its centre; the marginal belt alone increases in thickness. The piece thus becomes a bowl-shaped porringer, surrounded by a wide, flat edge.

The time for the laying has come. With one quick emission, the viscous, pale-yellow eggs are laid in the basin, where they heap together in the shape of a globe which projects largely outside the cavity. The spinnerets are once more set going. With short movements, as the tip of the abdomen rises and falls to weave the round mat, they cover up the exposed hemisphere. The result is a pill set in the middle of a circular carpet.

The legs, hitherto idle, are now working. They take up and break off one by one the threads that keep the round mat stretched on the coarse supporting network. At the same time the fangs grip this sheet, lift it by degrees, tear it from its base and fold it over upon the globe of eggs. It is a laborious operation. The whole edifice totters, the floor collapses, fouled with sand. By a movement of the legs, those soiled shreds are cast aside. Briefly, by means of violent tugs of the fangs, which pull, and broom-like efforts of the legs, which clear away, the Lycosa extricates the bag of eggs and removes it as a clear-cut mass, free from any adhesion.

It is a white-silk pill, soft to the touch and glutinous. Its size is that of an average cherry. An observant eye will notice, running horizontally around the middle, a fold which a needle is able to raise without breaking it. This hem, generally undistinguishable from the rest of the surface, is none other than the edge of the circular mat, drawn over the lower hemisphere. The other hemisphere, through which the youngsters will go out, is less well fortified: its only wrapper is the texture spun over the eggs immediately after they were laid.

The work of spinning, followed by that of tearing, is continued for a whole morning, from five to nine o'clock. Worn out with fatigue, the mother embraces her dear pill and remains motionless. I shall see no more to-day. Next morning, I find the Spider carrying the bag of eggs slung from her stern.

Henceforth, until the hatching, she does not leave go of the precious burden, which, fastened to the spinnerets by a short ligament, drags and bumps along the ground. With this load banging against her heels, she goes about her business; she walks or rests, she seeks her prey, attacks it and devours it. Should some accident cause the wallet to drop off, it is soon replaced. The spinnerets touch it somewhere, anywhere, and that is enough: adhesion is at once restored.

When the work is done, some of them emancipate themselves, think they will have a look at the country before retiring for good and all. It is these whom we meet at times, wandering aimlessly and dragging their bag behind them. Sooner or later, however, the vagrants return home; and the month of August is not over before a straw rustled in any burrow will bring the mother up, with her wallet slung behind her. I am able to procure as many as I want and, with them, to indulge in certain experiments of the highest interest.

It is a sight worth seeing, that of the Lycosa dragging her treasure after her, never leaving it, day or night, sleeping or waking, and defending it with a courage that strikes the beholder with awe. If I try to take the bag from her, she presses it to her breast in despair, hangs on to my pincers, bites them with her poison-fangs. I can hear the daggers grating on the steel. No, she would not allow herself to be robbed of the wallet with impunity, if my fingers were not supplied with an implement.

By dint of pulling and shaking the pill with the forceps, I take it from the Lycosa, who protests furiously. I fling her in exchange a pill taken from another Lycosa. It is at once seized in the fangs, embraced by the legs and hung on to the spinneret. Her own or another's: it is all one to the Spider, who walks away proudly with the alien wallet. This was to be expected, in view of the similarity of the pills exchanged.

A test of another kind, with a second subject, renders the mistake more striking. I substitute, in the place of the lawful bag which I have removed, the work of the Silky Epeira. The colour and softness of the material are the same in both cases; but the shape is quite different. The stolen object is a globe; the object presented in exchange is an elliptical conoid studded with angular projections along the edge of the base. The Spider takes no account of this dissimilarity. She promptly glues the queer bag to her spinnerets and is as pleased as though she were in possession of her real pill. My experimental villainies have no other consequence beyond an ephemeral carting. When hatching-time arrives, early in the case of Lycosa, late in that of the Epeira, the gulled Spider abandons the strange bag and pays it no further attention.

Let us penetrate yet deeper into the wallet-bearer's stupidity. After depriving the Lycosa of her eggs, I throw her a ball of cork, roughly polished with a file and of the same size as the stolen pill. She accepts the corky substance, so different from the silk purse, without the least demur. One would have thought that she would recognize her mistake with those eight eyes of hers, which gleam like precious stones. The silly creature pays no attention. Lovingly she embraces the cork ball, fondles it with her palpi, fastens it to her spinnerets and thenceforth drags it after her as though she were dragging her own bag.

Let us give another the choice between the imitation and the real. The rightful pill and the cork ball are placed together on the floor of the jar. Will the Spider be able to know the one that belongs to her? The fool is incapable of doing so. She makes a wild rush and seizes haphazard at one time her property, at another my sham product. Whatever is first touched becomes a good capture and is forthwith hung up.

If I increase the number of cork balls, if I put in four or five of them, with the real pill among them, it is seldom that the Lycosa recovers her own property. Attempts at inquiry, attempts at selection there are none. Whatever she snaps up at random she sticks to, be it good or bad. As there are more of the sham pills of cork, these are the most often seized by the Spider.

This obtuseness baffles me. Can the animal be deceived by the soft contact of the cork? I replace the cork balls by pellets of cotton or paper, kept in their round shape with a few bands of thread. Both are very readily accepted instead of the real bag that has been removed.

Can the illusion be due to the colouring, which is light in the cork and not unlike the tint of the silk globe when soiled with a little earth, while it is white in the

paper and the cotton, when it is identical with that of the original pill? I give the
Lycosa, in exchange for her work, a pellet of silk thread, chosen of a fine red, the
brightest of all colours. The uncommon pill is as readily accepted and as jealously guarded as the others.

THE FAMILY.

For three weeks and more the Lycosa trails the bag of eggs hanging to her spinnerets. The reader will remember the experiments described in the preceding section, particularly those with the cork ball and the thread pellet which the Spider
so foolishly accepts in exchange for the real pill. Well, this exceedingly dull-wit-
ted mother, satisfied with aught that knocks against her heels, is about to make us
wonder at her devotion.

Whether she come up from her shaft to lean upon the kerb and bask in the sun,
whether she suddenly retire underground in the face of danger, or whether she be
roaming the country before settling down, never does she let go her precious bag,
that very cumbrous burden in walking, climbing or leaping. If, by some accident,
it become detached from the fastening to which it is hung, she flings herself
madly on her treasure and lovingly embraces it, ready to bite whoso would take
it from her. I myself am sometimes the thief. I then hear the points of the poison-
fangs grinding against the steel of my pincers, which tug in one direction while
the Lycosa tugs in the other. But let us leave the animal alone: with a quick touch
of the spinnerets, the pill is restored to its place; and the Spider strides off, still
menacing.

Towards the end of summer, all the householders, old or young, whether in captivity on the window-sill or at liberty in the paths of the enclosure, supply me
daily with the following improving sight. In the morning, as soon as the sun is
hot and beats upon their burrow, the anchorites come up from the bottom with
their bag and station themselves at the opening. Long siestas on the threshold in
the sun are the order of the day throughout the fine season; but, at the present
time, the position adopted is a different one. Formerly, the Lycosa came out into
the sun for her own sake. Leaning on the parapet, she had the front half of her
body outside the pit and the hinder half inside. The eyes took their fill of light;
the belly remained in the dark. When carrying her egg-bag, the Spider reverses the
posture: the front is in the pit, the rear outside. With her hind-legs she holds the
white pill bulging with germs lifted above the entrance; gently she turns and turns
it, so as to present every side to the life-giving rays. And this goes on for half the
day, so long as the temperature is high; and it is repeated daily, with exquisite
patience, during three or four weeks. To hatch its eggs, the bird covers them with

the quilt of its breast; it strains them to the furnace of its heart. The Lycosa turns hers in front of the hearth of hearths: she gives them the sun as an incubator.

In the early days of September the young ones, who have been some time hatched, are ready to come out.

The whole family emerges from the bag straightway. Then and there, the young-sters climb to the mother's back. As for the empty bag, now a worthless shred, it is flung out of the burrow; the Lycosa does not give it a further thought. Huddled together, sometimes in two or three layers, according to their number, the little ones cover the whole back of the mother, who, for seven or eight months to come, will carry her family night and day. Nowhere can we hope to see a more edifying domestic picture than that of the Lycosa clothed in her young.

From time to time I meet a little band of gipsies passing along the high-road on their way to some neighbouring fair. The new-born babe mewls on the mother's breast, in a hammock formed out of a kerchief. The last-weaned is carried pick-a-back; a third toddles clinging to its mother's skirts; others follow closely, the biggest in the rear, ferreting in the blackberry-laden hedgerows. It is a magnificent spectacle of happy-go-lucky fruitfulness. They go their way, penniless and rejoic-ing. The sun is hot and the earth is fertile.

But how this picture pales before that of the Lycosa, that incomparable gipsy whose brats are numbered by the hundred! And one and all of them, from September to April, without a moment's respite, find room upon the patient crea-ture's back, where they are content to lead a tranquil life and to be carted about.

The little ones are very good; none moves, none seeks a quarrel with his neigh-bours. Clinging together, they form a continuous drapery, a shaggy ulster under which the mother becomes unrecognizable. Is it an animal, a fluff of wool, a clus-ter of small seeds fastened to one another? 'Tis impossible to tell at the first glance.

The equilibrium of this living blanket is not so firm but that falls often occur, especially when the mother climbs from indoors and comes to the threshold to let the little ones take the sun. The least brush against the gallery unseats a part of the family. The mishap is not serious. The Hen, fidgeting about her Chicks, looks for the strays, calls them, gathers them together. The Lycosa knows not these maternal alarms. Impassively, she leaves those who drop off to manage their own difficulty, which they do with wonderful quickness. Commend me to those youngsters for getting up without whining, dusting themselves and resuming their seat in the saddle! The unhorsed ones promptly find a leg of the mother, the usual climbing-pole; they swarm up it as fast as they can and recover their places on the bearer's back. The living bark of animals is reconstructed in the twinkling of an eye.

To speak here of mother-love were, I think, extravagant. The Lycosa's affection for her offspring hardly surpasses that of the plant, which is unacquainted with any tender feeling and nevertheless bestows the nicest and most delicate care upon its seeds. The animal, in many cases, knows no other sense of motherhood. What cares the Lycosa for her brood! She accepts another's as readily as her own; she is satisfied so long as her back is burdened with a swarming crowd, whether it issue from her ovaries or elsewhere. There is no question here of real maternal affection.

I have described elsewhere the prowess of the Copris watching over cells that are not her handiwork and do not contain her offspring. With a zeal which even the additional labour laid upon her does not easily weary, she removes the mildew from the alien dung-balls, which far exceed the regular nests in number; she gently scrapes and polishes and repairs them; she listens attentively and enquires by ear into each nurseling's progress. Her real collection could not receive greater care. Her own family or another's: it is all one to her.

The Lycosa is equally indifferent. I take a hair-pencil and sweep the living burden from one of my Spiders, making it fall close to another covered with her little ones. The evicted youngsters scamper about, find the new mother's legs outspread, nimbly clamber up these and mount on the back of the obliging creature, who quietly lets them have their way. They slip in among the others, or, when the layer is too thick, push to the front and pass from the abdomen to the thorax and even to the head, though leaving the region of the eyes uncovered. It does not do to blind the bearer: the common safety demands that. They know this and respect the lenses of the eyes, however populous the assembly be. The whole animal is now covered with a swarming carpet of young, all except the legs, which must preserve their freedom of action, and the under part of the body, where contact with the ground is to be feared.

My pencil forces a third family upon the already over-burdened Spider; and this too is peacefully accepted. The youngsters huddle up closer, lie one on top of the other in layers and room is found for all. The Lycosa has lost the last semblance of an animal, has become a nameless bristling thing that walks about. Falls are frequent and are followed by continual climbings.

I perceive that I have reached the limits, not of the bearer's good-will, but of equilibrium. The Spider would adopt an indefinite further number of foundlings, if the dimensions of her back afforded them a firm hold. Let us be content with this. Let us restore each family to its mother, drawing at random from the lot. There must necessarily be interchanges, but that is of no importance: real children and adopted children are the same thing in the Lycosa's eyes.

One would like to know if, apart from my artifices, in circumstances where I do not interfere, the good-natured dry-nurse sometimes burdens herself with a supplementary family; it would also be interesting to learn what comes of this association of lawful offspring and strangers. I have ample materials wherewith to obtain an answer to both questions. I have housed in the same cage two elderly matrons laden with youngsters. Each has her home as far removed from the other's as the size of the common pan permits. The distance is nine inches or more. It is not enough. Proximity soon kindles fierce jealousies between those intolerant creatures, who are obliged to live far apart so as to secure adequate hunting-grounds.

One morning I catch the two harridans fighting out their quarrel on the floor. The loser is laid flat upon her back; the victress, belly to belly with her adversary, clutches her with her legs and prevents her from moving a limb. Both have their poison-fangs wide open, ready to bite without yet daring, so mutually formidable are they. After a certain period of waiting, during which the pair merely exchange threats, the stronger of the two, the one on top, closes her lethal engine and grinds the head of the prostrate foe. Then she calmly devours the deceased by small mouthfuls.

Now what do the youngsters do, while their mother is being eaten? Easily consoled, heedless of the atrocious scene, they climb on the conqueror's back and quietly take their places among the lawful family. The ogress raises no objection, accepts them as her own. She makes a meal off the mother and adopts the orphans.

Let us add that, for many months yet, until the final emancipation comes, she will carry them without drawing any distinction between them and her own young. Henceforth the two families, united in so tragic a fashion, will form but one. We see how greatly out of place it would be to speak, in this connection, of mother-love and its fond manifestations.

Does the Lycosa at least feed the younglings who, for seven months, swarm upon her back? Does she invite them to the banquet when she has secured a prize? I thought so at first; and, anxious to assist at the family repast, I devoted special attention to watching the mothers eat. As a rule, the prey is consumed out of sight, in the burrow; but sometimes also a meal is taken on the threshold, in the open air. Besides, it is easy to rear the Lycosa and her family in a wire-gauze cage, with a layer of earth wherein the captive will never dream of sinking a well, such work being out of season. Everything then happens in the open.

Well, while the mother munches, chews, expresses the juices and swallows, the youngsters do not budge from their camping-ground on her back. Not one quits its place nor gives a sign of wishing to slip down and join in the meal. Nor does the mother extend an invitation to them to come and recruit themselves, nor put any broken victuals aside for them. She feeds and the others look on, or rather remain indifferent to what is happening. Their perfect quiet during the Lycosa's feast points to the possession of a stomach that knows no cravings.

Then with what are they sustained, during their seven months' upbringing on the mother's back? One conceives a notion of exudations supplied by the bearer's body, in which case the young would feed on their mother, after the manner of parasitic vermin, and gradually drain her strength.

We must abandon this notion. Never are they seen to put their mouths to the skin that should be a sort of teat to them. On the other hand, the Lycosa, far from being exhausted and shrivelling, keeps perfectly well and plump. She has the same pot-belly when she finishes rearing her young as when she began. She has not lost weight: far from it; on the contrary, she has put on flesh: she has gained the wherewithal to beget a new family next summer, one as numerous as to-day's.

Once more, with what do the little ones keep up their strength? We do not like to suggest reserves supplied by the egg as rectifying the animal's expenditure of vital force, especially when we consider that those reserves, themselves so close to nothing, must be economized in view of the silk, a material of the highest importance, of which a plentiful use will be made presently. There must be other powers at play in the tiny animal's machinery.

Total abstinence from food could be understood, if it were accompanied by inertia: immobility is not life. But the young Lycosae, though usually quiet on their mother's back, are at all times ready for exercise and for agile swarming. When they fall from the maternal perambulator, they briskly pick themselves up, briskly scramble up a leg and make their way to the top. It is a splendidly nimble and spirited performance. Besides, once seated, they have to keep a firm balance in the mass; they have to stretch and stiffen their little limbs in order to hang on to their neighbours. As a matter of fact, there is no absolute rest for them. Now physiology teaches us that not a fibre works without some expenditure of energy. The animal, which can be likened, in no small measure, to our industrial machines, demands, on the one hand, the renovation of its organism, which wears out with movement, and, on the other, the maintenance of the heat transformed into action. We can compare it with the locomotive-engine. As the iron horse performs its work, it gradually wears out its pistons, its rods, its wheels, its boiler-tubes, all of which have to be made good from time to time. The founder and the smith repair it, supply it, so to speak, with 'plastic food,' the food that becomes embodied with the whole and forms part of it. But, though it have just come from the engine-shop, it is still inert. To acquire the power of movement it must receive from the stoker a supply of 'energy-producing food'; in other words, he lights a few shovelfuls of coal in its inside. This heat will produce mechanical work.

Even so with the beast. As nothing is made from nothing, the egg supplies first the materials of the new-born animal; then the plastic food, the smith of living creatures, increases the body, up to a certain limit, and renews it as it wears away. The stoker works at the same time, without stopping. Fuel, the source of energy, makes but a short stay in the system, where it is consumed and furnishes heat, whence movement is derived. Life is a fire-box. Warmed by its food, the animal machine moves, walks, runs, jumps, swims, flies, sets its locomotory apparatus going in a thousand manners.

To return to the young Lycosae, they grow no larger until the period of their emancipation. I find them at the age of seven months the same as when I saw them at their birth. The egg supplied the materials necessary for their tiny frames; and, as the loss of waste substance is, for the moment, excessively small, or even nil, additional plastic food is not needed so long as the wee creature does not grow. In this respect, the prolonged abstinence presents no difficulty. But there remains the question of energy-producing food, which is indispensable, for the little Lycosa moves, when necessary, and very actively at that. To what shall we attribute the heat expended upon action, when the animal takes absolutely no nourishment?

An idea suggests itself. We say to ourselves that, without being life, a machine is something more than matter, for man has added a little of his mind to it. Now the iron beast, consuming its ration of coal, is really browsing the ancient foliage of arborescent ferns in which solar energy has accumulated.

Beasts of flesh and blood act no otherwise. Whether they mutually devour one another or levy tribute on the plant, they invariably quicken themselves with the stimulant of the sun's heat, a heat stored in grass, fruit, seed and those which feed on such. The sun, the soul of the universe, is the supreme dispenser of energy.

Instead of being served up through the intermediary of food and passing through the ignominious circuit of gastric chemistry, could not this solar energy penetrate the animal directly and charge it with activity, even as the battery charges an accumulator with power? Why not live on sun, seeing that, after all, we find naught but sun in the fruits which we consume?

Chemical science, that bold revolutionary, promises to provide us with synthetic foodstuffs. The laboratory and the factory will take the place of the farm. Why should not physical science step in as well? It would leave the preparation of plastic food to the chemist's retorts; it would reserve for itself that of energy-producing food which, reduced to its exact terms, ceases to be matter. With the aid of some ingenious apparatus, it would pump into us our daily ration of solar energy, to be later expended in movement, whereby the machine would be kept going without the often painful assistance of the stomach and its adjuncts. What a delightful world, where one could lunch off a ray of sunshine!

Is it a dream, or the anticipation of a remote reality? The problem is one of the most important that science can set us. Let us first hear the evidence of the young Lycosae regarding its possibilities.

For seven months, without any material nourishment, they expend strength in moving. To wind up the mechanism of their muscles, they recruit themselves direct with heat and light. During the time when she was dragging the bag of eggs behind her, the mother, at the best moments of the day, came and held up her pill to the sun. With her two hind-legs she lifted it out of the ground into the full light; slowly she turned it and turned it, so that every side might receive its share of the vivifying rays. Well, this bath of life, which awakened the germs, is now prolonged to keep the tender babes active.

Daily, if the sky be clear, the Lycosa, carrying her young, comes up from the burrow, leans on the kerb and spends long hours basking in the sun. Here, on their mother's back, the youngsters stretch their limbs delightedly, saturate themselves with heat, take in reserves of motor-power, absorb energy.

They are motionless; but, if I only blow upon them, they stampede as nimbly as though a hurricane were passing. Hurriedly, they disperse; hurriedly, they reassemble: a proof that, without material nourishment, the little animal machine is always at full pressure, ready to work. When the shade comes, mother and sons go down again, surfeited with solar emanations. The feast of energy at the Sun Tavern is finished for the day.

CHAPTER 10.

THE BANDED EPEIRA.

BUILDING THE WEB.

The fowling-snare is one of man's ingenious villainies. With lines, pegs and poles, two large, earth-coloured nets are stretched upon the ground, one to the right, the other to the left of a bare surface. A long cord, pulled at the right moment by the fowler, who hides in a brushwood hut, works them and brings them together suddenly, like a pair of shutters.

Divided between the two nets are the cages of the decoy-birds—Linnets and Chaffinches, Greenfinches and Yellowhammers, Buntings and Ortolans—sharp-eared creatures which, on perceiving the distant passage of a flock of their own kind, forthwith utter a short calling note. One of them, the Sambé, an irresistible tempter, hops about and flaps his wings in apparent freedom. A bit of twine fastens him to his convict's stake. When, worn with fatigue and driven desperate by his vain attempts to get away, the sufferer lies down flat and refuses to do his duty, the fowler is able to stimulate him without stirring from his hut. A long string sets in motion a little lever working on a pivot. Raised from the ground by this diabolical contrivance, the bird flies, falls down and flies up again at each jerk of the cord.

The fowler waits, in the mild sunlight of the autumn morning. Suddenly, great excitement in the cages. The Chaffinches chirp their rallying cry:

"Pinck! Pinck!"

There is something happening in the sky. The Sambé, quick! They are coming, the simpletons; they swoop down upon the treacherous floor. With a rapid movement, the man in ambush pulls his string. The nets close and the whole flock is caught.

Man has wild beast's blood in his veins. The fowler hastens to the slaughter. With his thumb he stifles the beating of the captives' hearts, staves in their skulls. The little birds, so many piteous heads of game, will go to market, strung in dozens on a wire passed through their nostrils.

For scoundrelly ingenuity, the Epeira's net can bear comparison with the fowler's; it even surpasses it when, on patient study, the main features of its supreme perfection stand revealed. What refinement of art for a mess of Flies! Nowhere, in the whole animal kingdom, has the need to eat inspired a more cunning industry. If the reader will meditate upon the description that follows, he will certainly share my admiration.

In bearing and colouring, Epeira fasciata is the handsomest of the Spiders of the South. On her fat belly, a mighty silk-warehouse nearly as large as a hazel-nut, are alternate yellow, black and silver sashes, to which she owes her epithet of Banded. Around that portly abdomen the eight long legs, with their dark- and pale-brown rings, radiate like spokes.

Any small prey suits her; and, as long as she can find supports for her web, she settles wherever the Locust hops, wherever the Fly hovers, wherever the Dragon-fly dances or the Butterfly flits. As a rule, because of the greater abundance of game, she spreads her toils across some brooklet, from bank to bank among the rushes. She also stretches them, but not so assiduously, in the thickets of evergreen oak, on the slopes with the scrubby greenswards, dear to the Grasshoppers.

Her hunting-weapon is a large upright web, whose outer boundary, which varies according to the disposition of the ground, is fastened to the neighbouring branches by a number of moorings. Let us see, first of all, how the ropes which form the framework of the building are obtained.

All day invisible, crouching amid the cypress-leaves, the Spider, at about eight o'clock in the evening, solemnly emerges from her retreat and makes for the top of a branch. In this exalted position she sits for sometime laying her plans with due regard to the locality; she consults the weather, ascertains if the night will be fine. Then, suddenly, with her eight legs widespread, she lets herself drop straight down, hanging to the line that issues from her spinnerets. Just as the rope-maker obtains the even output of his hemp by walking backwards, so does the Epeira obtain the discharge of hers by falling. It is extracted by the weight of her body.

The descent, however, has not the brute speed which the force of gravity would give it, if uncontrolled. It is governed by the action of the spinnerets, which contract or expand their pores, or close them entirely, at the faller's pleasure. And so, with gentle moderation, she pays out this living plumb-line, of which my lantern clearly shows me the plumb, but not always the line. The great squab seems at such times to be sprawling in space, without the least support.

She comes to an abrupt stop two inches from the ground; the silk-reel ceases working. The Spider turns round, clutches the line which she has just obtained and climbs up by this road, still spinning. But, this time, as she is no longer assisted by the force of gravity, the thread is extracted in another manner. The two hind-legs, with a quick alternate action, draw it from the wallet and let it go.

On returning to her starting-point, at a height of six feet or more, the Spider is now in possession of a double line, bent into a loop and floating loosely in a current of air. She fixes her end where it suits her and waits until the other end, wafted by the wind, has fastened its loop to the adjacent twigs.

Feeling her thread fixed, the Epeira runs along it repeatedly, from end to end, adding a fibre to it on each journey. Whether I help or not, this forms the "suspension cable," the main piece of the framework. I call it a cable, in spite of its extreme thinness, because of its structure. It looks as though it were single, but, at the two ends, it is seen to divide and spread, tuft-wise, into numerous constituent parts, which are the product of as many crossings. These diverging fibres, with their several contact-points, increase the steadiness of the two extremities.

The suspension-cable is incomparably stronger than the rest of the work and lasts for an indefinite time. The web is generally shattered after the night's hunting and is nearly always rewoven on the following evening. After the removal of the wreckage, it is made all over again, on the same site, cleared of everything except the cable from which the new network is to hang.

Once the cable is laid, in this way or in that, the Spider is in possession of a base that allows her to approach or withdraw from the leafy piers at will. From the height of the cable she lets herself slip to a slight depth, varying the points of her fall. In this way she obtains, to right and left, a few slanting cross-bars, connecting the cable with the branches.

These cross-bars, in their turn, support others in ever changing directions. When there are enough of them, the Epeira need no longer resort to falls in order to extract her threads; she goes from one cord to the next, always wire-drawing with her hind-legs. This results in a combination of straight lines owning no order, save that they are kept in one nearly perpendicular plane. Thus is marked out a very irregular polygonal area, wherein the web, itself a work of magnificent regularity, shall presently be woven.

In the lower part of the web, starting from the centre, a wide opaque ribbon descends zigzag-wise across the radii. This is the Epeira's trade-mark, the flourish of an artist initialling his creation. "Fecit So-and-so," she seems to say, when giving the last throw of the shuttle to her handiwork.

That the Spider feels satisfied when, after passing and repassing from spoke to spoke, she finishes her spiral, is beyond a doubt: the work achieved ensures her food for a few days to come. But, in this particular case, the vanity of the spin-stress has naught to say to the matter: the strong silk zigzag is added to impart greater firmness to the web.

THE LIME-SNARE.

The spiral network of the Epeirae possesses contrivances of fearsome cunning. The thread that forms it is seen with the naked eye to differ from that of the framework and the spokes. It glitters in the sun, looks as though it were knotted and gives the impression of a chaplet of atoms. To examine it through the lens on the web itself is scarcely feasible, because of the shaking of the fabric, which trembles at the least breath. By passing a sheet of glass under the web and lifting it, I take away a few pieces of thread to study, pieces that remain fixed to the glass in parallel lines. Lens and microscope can now play their part.

The sight is perfectly astounding. Those threads, on the borderland between the visible and the invisible, are very closely twisted twine, similar to the gold cord of our officers' sword-knots. Moreover, they are hollow. The infinitely slender is a tube, a channel full of a viscous moisture resembling a strong solution of gum arabic. I can see a diaphanous trail of this moisture trickling through the broken ends. Under the pressure of the thin glass slide that covers them on the stage of the microscope, the twists lengthen out, become crinkled ribbons, traversed from end to end, through the middle, by a dark streak, which is the empty container.

The fluid contents must ooze slowly through the side of those tubular threads, rolled into twisted strings, and thus render the network sticky. It is sticky, in fact, and in such a way as to provoke surprise. I bring a fine straw flat down upon three or four rungs of a sector. However gentle the contact, adhesion is at once established. When I lift the straw, the threads come with it and stretch to twice or three times their length, like a thread of india-rubber. At last, when over-taut, they loosen without breaking and resume their original form. They lengthen by unrolling their twist, they shorten by rolling it again; lastly, they become adhesive by taking the glaze of the gummy moisture wherewith they are filled.

In short, the spiral thread is a capillary tube finer than any that our physics will ever know. It is rolled into a twist so as to possess an elasticity that allows it, without breaking, to yield to the tugs of the captured prey; it holds a supply of sticky matter in reserve in its tube, so as to renew the adhesive properties of the surface by incessant exudation, as they become impaired by exposure to the air. It is simply marvellous.

The Epeira hunts not with springs, but with lime-snares. And such lime-snares! Everything is caught in them, down to the dandelion-plume that barely brushes against them. Nevertheless, the Epeira, who is in constant touch with her web, is not caught in them. Why? Because the Spider has contrived for herself, in the middle of her trap, a floor in whose construction the sticky spiral thread plays no part. There is here, covering a space which, in the larger webs, is about equal to the palm of one's hand, a neutral fabric in which the exploring straw finds no adhesiveness anywhere.

Here, on this central resting-floor, and here only, the Epeira takes her stand, waiting whole days for the arrival of the game. However close, however prolonged her contact with this portion of the web, she runs no risk of sticking to it, because the gummy coating is lacking, as is the twisted and tubular structure, throughout the length of the spokes and throughout the extent of the auxiliary spiral. These pieces, together with the rest of the framework, are made of plain, straight, solid thread.

But when a victim is caught, sometimes right at the edge of the web, the Spider has to rush up quickly, to bind it and overcome its attempts to free itself. She is walking then upon her network; and I do not find that she suffers the least inconvenience. The lime-threads are not even lifted by the movements of her legs.

In my boyhood, when a troop of us would go, on Thursdays (The weekly half-day in French schools.—Translator's Note.), to try and catch a Goldfinch in the hemp-fields, we used, before covering the twigs with glue, to grease our fingers with a few drops of oil, lest we should get them caught in the sticky matter. Does the Epeira know the secret of fatty substances? Let us try.

I rub my exploring straw with slightly oiled paper. When applied to the spiral thread of the web, it now no longer sticks to it. The principle is discovered. I pull out the leg of a live Epeira. Brought just as it is into contact with the lime-threads, it does not stick to them any more than to the neutral cords, whether spokes or part of the framework. We were entitled to expect this, judging by the Spider's general immunity.

But here is something that wholly alters the result. I put the leg to soak for a quarter of an hour in disulphide of carbon, the best solvent of fatty matters. I wash it carefully with a brush dipped in the same fluid. When this washing is finished, the leg sticks to the snaring-thread quite easily and adheres to it just as well as anything else would, the unoiled straw, for instance.

Did I guess aright when I judged that it was a fatty substance that preserved the Epeira from the snares of her sticky Catherine-wheel? The action of the carbon-disulphide seems to say yes. Besides, there is no reason why a substance of this kind, which plays so frequent a part in animal economy, should not coat the Spider very slightly by the mere act of perspiration. We used to rub our fingers with a little oil before handling the twigs in which the Goldfinch was to be caught; even so the Epeira varnishes herself with a special sweat, to operate on any part of her web without fear of the lime-threads.

However, an unduly protracted stay on the sticky threads would have its drawbacks. In the long run, continual contact with those threads might produce a certain adhesion and inconvenience to the Spider, who must preserve all her agility in order to rush upon the prey before it can release itself. For this reason, gummy threads are never used in building the post of interminable waiting.

It is only on her resting-floor that the Epeira sits, motionless and with her eight legs outspread, ready to mark the least quiver in the net. It is here, again, that she takes her meals, often long-drawn out, when the joint is a substantial one; it is hither that, after trussing and nibbling it, she drags her prey at the end of a thread, to consume it at her ease on a non-viscous mat. As a hunting-post and refectory, the Epeira has contrived a central space, free from glue.

As for the glue itself, it is hardly possible to study its chemical properties, because the quantity is so slight. The microscope shows it trickling from the broken threads in the form of a transparent and more or less granular streak. The following experiment will tell us more about it.

With a sheet of glass passed across the web, I gather a series of lime-threads which remain fixed in parallel lines. I cover this sheet with a bell-jar standing in a depth of water. Soon, in this atmosphere saturated with humidity, the threads become enveloped in a watery sheath, which gradually increases and begins to flow. The twisted shape has by this time disappeared; and the channel of the thread reveals a chaplet of translucent orbs, that is to say, a series of extremely fine drops.

In twenty-four hours the threads have lost their contents and are reduced to almost invisible streaks. If I then lay a drop of water on the glass, I get a sticky solution similar to that which a particle of gum arabic might yield. The conclusion is evident: the Epeira's glue is a substance that absorbs moisture freely. In an atmosphere with a high degree of humidity, it becomes saturated and percolates by sweating through the side of the tubular threads.

These data explain certain facts relating to the work of the net. The Epeirae weave at very early hours, long before dawn. Should the air turn misty, they sometimes leave that part of the task unfinished: they build the general framework, they lay the spokes, they even draw the auxiliary spiral, for all these parts are unaffected by excess of moisture; but they are very careful not to work at the lime-threads, which, if soaked by the fog, would dissolve into sticky shreds and lose their efficacy by being wetted. The net that was started will be finished to-morrow, if the atmosphere be favourable.

While the highly-absorbent character of the snaring-thread has its drawbacks, it also has compensating advantages. The Epeirae, when hunting by day, affect those hot places, exposed to the fierce rays of the sun, wherein the Crickets delight. In the torrid heats of the dog-days, therefore, the lime-threads, but for special provisions, would be liable to dry up, to shrivel into stiff and lifeless filaments. But the very opposite happens. At the most scorching times of the day they continue supple, elastic and more and more adhesive.

How is this brought about? By their very powers of absorption. The moisture of which the air is never deprived penetrates them slowly; it dilutes the thick contents of their tubes to the requisite degree and causes it to ooze through, as and when the earlier stickiness decreases. What bird-catcher could vie with the Garden Spider in the art of laying lime-snares? And all this industry and cunning for the capture of a Moth!

I should like an anatomist endowed with better implements than mine and with less tired eyesight to explain to us the work of the marvellous rope-yard. How is the silken matter moulded into a capillary tube? How is this tube filled with glue and tightly twisted? And how does this same mill also turn out plain threads, wrought first into a framework and then into muslin and satin? What a number of products to come from that curious factory, a Spider's belly! I behold the results, but fail to understand the working of the machine. I leave the problem to the masters of the microtome and the scalpel.

THE HUNT.

The Epeirae are monuments of patience in their lime-snare. With her head down and her eight legs widespread, the Spider occupies the centre of the web, the receiving-point of the information sent along the spokes. If anywhere, behind or before, a vibration occur, the sign of a capture, the Epeira knows about it, even without the aid of sight. She hastens up at once.

Until then, not a movement: one would think that the animal was hypnotized by her watching. At most, on the appearance of anything suspicious, she begins shaking her nest. This is her way of inspiring the intruder with awe. If I myself wish to provoke the singular alarm, I have but to tease the Epeira with a bit of straw. You cannot have a swing without an impulse of some sort. The terror-stricken Spider, who wishes to strike terror into others, has hit upon something much better. With nothing to push her, she swings with the floor of ropes. There is no effort, no visible exertion. Not a single part of the animal moves; and yet everything trembles. Violent shaking proceeds from apparent inertia. Rest causes commotion.

When calm is restored, she resumes her attitude, ceaselessly pondering the harsh problem of life:

"Shall I dine to-day, or not?"

Certain privileged beings, exempt from those anxieties, have food in abundance and need not struggle to obtain it. Such is the Gentle, who swims blissfully in the broth of the putrefying Adder. Others—and, by a strange irony of fate, these are generally the most gifted—only manage to eat by dint of craft and patience.

You are of their company, O my industrious Epeirae! So that you may dine, you spend your treasures of patience nightly; and often without result. I sympathize with your woes, for I, who am as concerned as you about my daily bread, I also doggedly spread my net, the net for catching ideas, a more elusive and less substantial prize than the Moth. Let us not lose heart. The best part of life is not in the present, still less in the past; it lies in the future, the domain of hope. Let us wait.

All day long, the sky, of a uniform grey, has appeared to be brewing a storm. In spite of the threatened downpour, my neighbour, who is a shrewd weather-prophet, has come out of the cypress-tree and begun to renew her web at the regular hour. Her forecast is correct: it will be a fine night. See, the steaming-pan of the clouds splits open; and, through the apertures, the moon peeps, inquisitively. I too, lantern in hand, am peeping. A gust of wind from the north clears the realms on high; the sky becomes magnificent; perfect calm reigns below. The Moths begin their nightly rounds. Good! One is caught, a mighty fine one. The Spider will dine to-day.

What happens next, in an uncertain light, does not lend itself to accurate observation. It is better to turn to those Garden Spiders who never leave their web and who hunt mainly in the daytime. The Banded and the Silky Epeira, both of whom live on the rosemaries in the enclosure, shall show us in broad daylight the innermost details of the tragedy.

I myself place on the lime-snare a victim of my selecting. Its six legs are caught without more ado. If the insect raises one of its tarsi and pulls towards itself, the treacherous thread follows, unwinds slightly and, without letting go or breaking, yields to the captive's desperate jerks. Any limb released only tangles the others still more and is speedily recaptured by the sticky matter. There is no means of escape, except by smashing the trap with a sudden effort whereof even powerful insects are not always capable.

Warned by the shaking of the net, the Epeira hastens up; she turns round about the quarry; she inspects it at a distance, so as to ascertain the extent of the danger before attacking. The strength of the snareling will decide the plan of campaign. Let us first suppose the usual case, that of an average head of game, a Moth or Fly of some sort. Facing her prisoner, the Spider contracts her abdomen slightly and touches the insect for a moment with the end of her spinnerets; then, with her front tarsi, she sets her victim spinning. The Squirrel, in the moving cylinder of his cage, does not display a more graceful or nimbler dexterity. A cross-bar of the sticky spiral serves as an axis for the tiny machine, which turns, turns swiftly, like a spit. It is a treat to the eyes to see it revolve.

What is the object of this circular motion? It is this: the brief contact of the spinnerets has given a starting-point for a thread, which the Spider must now draw from her silk warehouse and gradually roll around the captive, so as to swathe him in a winding-sheet which will overpower any effort made. It is the exact process employed in our wire-mills: a motor-driven spool revolves and, by its action, draws the wire through the narrow eyelet of a steel plate, making it of the fineness required, and, with the same movement, winds it round and round its collar.

Even so with the Epeira's work. The Spider's front tarsi are the motor; the revolving spool is the captured insect; the steel eyelet is the aperture of the spinnerets. To bind the subject with precision and dispatch nothing could be better than this inexpensive and highly effective method.

Less frequently, a second process is employed. With a quick movement, the Spider herself turns round about the motionless insect, crossing the web first at the top and then at the bottom and gradually placing the fastenings of her line. The great elasticity of the lime-threads allows the Epeira to fling herself time after time right into the web and to pass through it without damaging the net.

Let us now suppose the case of some dangerous game: a Praying Mantis, for instance, brandishing her lethal limbs, each hooked and fitted with a double saw; an angry Hornet, darting her awful sting; a sturdy Beetle, invincible under his horny armour. These are exceptional morsels, hardly ever known to the Epeirae. Will they be accepted, if supplied by my stratagems?

They are, but not without caution. The game is seen to be perilous of approach and the Spider turns her back upon it instead of facing it; she trains her rope-cannon upon it. Quickly the hind-legs draw from the spinnerets something much better than single cords. The whole silk-battery works at one and the same time, firing a regular volley of ribbons and sheets, which a wide movement of the legs spreads fan-wise and flings over the entangled prisoner. Guarding against sudden starts, the Epeira casts her armfuls of bands on the front- and hind-parts, over the legs and over the wings, here, there and everywhere, extravagantly. The most fiery prey is promptly mastered under this avalanche. In vain the Mantis tries to open her saw-toothed arm-guards; in vain the Hornet makes play with her dagger; in vain the Beetle stiffens his legs and arches his back: a fresh wave of threads swoops down and paralyses every effort.

The ancient retiarius, when pitted against a powerful wild beast, appeared in the arena with a rope-net folded over his left shoulder. The animal made its spring. The man, with a sudden movement of his right arm, cast the net after the manner of the fisherman; he covered the beast and tangled it in the meshes. A thrust of the trident gave the quietus to the vanquished foe.

The Epeira acts in like fashion, with this advantage, that she is able to renew her armful of fetters. Should the first not suffice, a second instantly follows and another and yet another, until the reserves of silk become exhausted.

When all movement ceases under the snowy winding-sheet, the Spider goes up to her bound prisoner. She has a better weapon than the bestiarius' trident: she has her poison-fangs. She gnaws at the Locust, without undue persistence, and then withdraws, leaving the torpid patient to pine away.

These lavished, far-flung ribbons threaten to exhaust the factory; it would be much more economical to resort to the method of the spool; but, to turn the machine, the Spider would have to go up to it and work it with her leg. This is too risky; and hence the continuous spray of silk, at a safe distance. When all is used up, there is more to come.

Still, the Epeira seems concerned at this excessive outlay. When circumstances permit, she gladly returns to the mechanism of the revolving spool. I saw her practice this abrupt change of tactics on a big Beetle, with a smooth, plump body, which lent itself admirably to the rotary process. After depriving the beast of all power of movement, she went up to it and turned her corpulent victim as she would have done with a medium-sized Moth.

But with the Praying Mantis, sticking out her long legs and her spreading wings, rotation is no longer feasible. Then, until the quarry is thoroughly subdued, the spray of bandages goes on continuously, even to the point of drying up the silk glands. A capture of this kind is ruinous. It is true that, except when I interfered, I have never seen the Spider tackle that formidable provender.

Be it feeble or strong, the game is now neatly trussed, by one of the two methods. The next move never varies. The bound insect is bitten, without persistency and without any wound that shows. The Spider next retires and allows the bite to act, which it soon does. She then returns.

If the victim be small, a Clothes-moth, for instance, it is consumed on the spot, at the place where it was captured. But, for a prize of some importance, on which she hopes to feast for many an hour, sometimes for many a day, the Spider needs a sequestered dining-room, where there is naught to fear from the stickiness of the network. Before going to it, she first makes her prey turn in the converse direction to that of the original rotation. Her object is to free the nearest spokes, which supplied pivots for the machinery. They are essential factors which it behoves her to keep intact, if need be by sacrificing a few cross-bars.

It is done; the twisted ends are put back into position. The well-trussed game is at last removed from the web and fastened on behind with a thread. The Spider then marches in front and the load is trundled across the web and hoisted to the resting-floor, which is both an inspection-post and a dining-hall. When the Spider is of a species that shuns the light and possesses a telegraph-line, she mounts to her daytime hiding-place along this line, with the game bumping against her heels.

While she is refreshing herself, let us enquire into the effects of the little bite previously administered to the silk-swathed captive. Does the Spider kill the patient with a view to avoiding unseasonable jerks, protests so disagreeable at dinner-time? Several reasons make me doubt it. In the first place, the attack is so much veiled as to have all the appearance of a mere kiss. Besides, it is made anywhere, at the first spot that offers. The expert slayers employ methods of the highest precision: they give a stab in the neck, or under the throat; they wound the cervical nerve-centres, the seat of energy. The paralysers, those accomplished anatomists, poison the motor nerve-centres, of which they know the number and position. The Epeira possesses none of this fearsome knowledge. She inserts her fangs at random, as the Bee does her sting. She does not select one spot rather than another; she bites indifferently at whatever comes within reach. This being so, her poison would have to possess unparalleled virulence to produce a corpse-like inertia no matter which the point attacked. I can scarcely believe in instantaneous death resulting from the bite, especially in the case of insects, with their highly-resistant organisms.

Besides, is it really a corpse that the Epeira wants, she who feeds on blood much more than on flesh? It were to her advantage to suck a live body, wherein the flow of the liquids, set in movement by the pulsation of the dorsal vessel, that rudimentary heart of insects, must act more freely than in a lifeless body, with its stagnant fluids. The game which the Spider means to suck dry might very well not be dead. This is easily ascertained.

I place some Locusts of different species on the webs in my menagerie, one on this, another on that. The Spider comes rushing up, binds the prey, nibbles at it gently and withdraws, waiting for the bite to take effect. I then take the insect and carefully strip it of its silken shroud. The Locust is not dead; far from it; one would even think that he had suffered no harm. I examine the released prisoner through the lens in vain; I can see no trace of a wound.

Can he be unscathed, in spite of the sort of kiss which I saw given to him just now? You would be ready to say so, judging by the furious way in which he kicks in my fingers. Nevertheless, when put on the ground, he walks awkwardly, he seems reluctant to hop. Perhaps it is a temporary trouble, caused by his terrible excitement in the web. It looks as though it would soon pass.

I lodge my Locusts in cages, with a lettuce-leaf to console them for their trials; but they will not be comforted. A day elapses, followed by a second. Not one of them touches the leaf of salad; their appetite has disappeared. Their movements become more uncertain, as though hampered by irresistible torpor. On the second day they are dead, everyone irrecoverably dead.

The Epeira, therefore, does not incontinently kill her prey with her delicate bite; she poisons it so as to produce a gradual weakness, which gives the blood-sucker ample time to drain her victim, without the least risk, before the rigor mortis stops the flow of moisture.

The meal lasts quite twenty-four hours, if the joint be large; and to the very end the butchered insect retains a remnant of life, a favourable condition for the exhausting of the juices. Once again, we see a skilful method of slaughter, very different from the tactics in use among the expert paralysers or slayers. Here there is no display of anatomical science. Unacquainted with the patient's structure, the Spider stabs at random. The virulence of the poison does the rest.

There are, however, some very few cases in which the bite is speedily mortal. My notes speak of an Angular Epeira grappling with the largest Dragon-fly in my district (Aeshna grandis, Lin.) I myself had entangled in the web this head of big game, which is not often captured by the Epeirae. The net shakes violently, seems bound to break its moorings. The Spider rushes from her leafy villa, runs boldly up to the giantess, flings a single bundle of ropes at her and, without further precautions, grips her with her legs, tries to subdue her and then digs her fangs into the Dragon-fly's back. The bite is prolonged in such a way as to astonish me. This is not the perfunctory kiss with which I am already familiar; it is a deep, determined wound. After striking her blow, the Spider retires to a certain distance and waits for her poison to take effect.

I at once remove the Dragon-fly. She is dead, really and truly dead. Laid upon my table and left alone for twenty-four hours, she makes not the slightest movement. A prick of which my lens cannot see the marks, so sharp-pointed are the Epeira's weapons, was enough, with a little insistence, to kill the powerful animal. Proportionately, the Rattlesnake, the Horned Viper, the Trigonocephalus and other ill-famed serpents produce less paralysing effects upon their victims.

And these Epeirae, so terrible to insects, I am able to handle without any fear. My skin does not suit them. If I persuaded them to bite me, what would happen to me? Hardly anything. We have more cause to dread the sting of a nettle than the dagger which is fatal to Dragon-flies. The same virus acts differently upon this organism and that, is formidable here and quite mild there. What kills the insect may easily be harmless to us. Let us not, however, generalize too far. The Narbonne Lycosa, that other enthusiastic insect-huntress, would make us pay dearly if we attempted to take liberties with her.

It is not uninteresting to watch the Epeira at dinner. I light upon one, the Banded Epeira, at the moment, about three o'clock in the afternoon, when she has captured a Locust. Planted in the centre of the web, on her resting-floor, she attacks the venison at the joint of a haunch. There is no movement, not even of the mouth-parts, so far as I am able to discover. The mouth lingers, close-applied, at the point originally bitten. There are no intermittent mouthfuls, with the mandibles moving backwards and forwards. It is a sort of continuous kiss.

I visit my Epeira at intervals. The mouth does not change its place. I visit her for the last time at nine o'clock in the evening. Matters stand exactly as they did: after six hours' consumption, the mouth is still sucking at the lower end of the right haunch. The fluid contents of the victim are transferred to the ogress's belly, I know not how.

Next morning, the Spider is still at table. I take away her dish. Naught remains of the Locust but his skin, hardly altered in shape, but utterly drained and perforated in several places. The method, therefore, was changed during the night. To extract the non-fluent residue, the viscera and muscles, the stiff cuticle had to be tapped here, there and elsewhere, after which the tattered husk, placed bodily in the press of the mandibles, would have been chewed, re-chewed and finally reduced to a pill, which the sated Spider throws up. This would have been the end of the victim, had I not taken it away before the time.

Whether she wound or kill, the Epeira bites her captive somewhere or other, no matter where. This is an excellent method on her part, because of the variety of the game that comes her way. I see her accepting with equal readiness whatever chance may send her: Butterflies and Dragon-flies, Flies and Wasps, small Dung-beetles and Locusts. If I offer her a Mantis, a Bumble-bee, an Anoxia—the equivalent of the common Cockchafer—and other dishes probably unknown to her race, she accepts all and any, large and small, thin-skinned and horny-skinned, that which goes afoot and that which takes winged flight. She is omnivorous, she preys on everything, down to her own kind, should the occasion offer.

Had she to operate according to individual structure, she would need an anatomical dictionary; and instinct is essentially unfamiliar with generalities: its knowledge is always confined to limited points. The Cerceres know their Weevils and their Buprestis-beetles absolutely; the Sphex their Grasshoppers, their Crickets and their Locusts; the Scoliae their Cetonia- and Oryctes-grubs. (The Scolia is a Digger-wasp, like the Cerceris and the Sphex, and feeds her larvae on the grubs of the Cetonia, or Rose-chafer, and the Oryctes, or Rhinoceros-beetle.— Translator's Note.) Even so the other paralysers. Each has her own victim and knows nothing of any of the others.

The same exclusive tastes prevail among the slayers. Let us remember, in this connection, Philanthus apivorus and, especially, the Thomisus, the comely Spider who cuts Bees' throats. They understand the fatal blow, either in the neck or under the chin, a thing which the Epeira does not understand; but, just because of this talent, they are specialists. Their province is the Domestic Bee.

Animals are a little like ourselves: they excel in an art only on condition of specializing in it. The Epeira, who, being omnivorous, is obliged to generalize, abandons scientific methods and makes up for this by distilling a poison capable of producing torpor and even death, no matter what the point attacked.

Recognizing the large variety of game, we wonder how the Epeira manages not to hesitate amid those many diverse forms, how, for instance, she passes from the Locust to the Butterfly, so different in appearance. To attribute to her as a guide an extensive zoological knowledge were wildly in excess of what we may reasonably expect of her poor intelligence. The thing moves, therefore it is worth catching: this formula seems to sum up the Spider's wisdom.

THE TELEGRAPH-WIRE.

Of the six Garden Spiders that form the object of my observations, two only, the Banded and the Silky Epeira, remain constantly in their webs, even under the blinding rays of a fierce sun. The others, as a rule, do not show themselves until nightfall. At some distance from the net they have a rough-and-ready retreat in the brambles, an ambush made of a few leaves held together by stretched threads. It is here that, for the most part, they remain in the daytime, motionless and sunk in meditation.

But the shrill light that vexes them is the joy of the fields. At such times the Locust hops more nimbly than ever, more gaily skims the Dragon-fly. Besides, the limy web, despite the rents suffered during the night, is still in serviceable condition. If some giddy-pate allow himself to be caught, will the Spider, at the distance whereto she has retired, be unable to take advantage of the windfall? Never fear. She arrives in a flash. How is she apprised? Let us explain the matter.

The alarm is given by the vibration of the web, much more than by the sight of the captured object. A very simple experiment will prove this. I lay upon a Banded Epeira's lime-threads a Locust that second asphyxiated with carbon disulphide. The carcass is placed in front, or behind, or at either side of the Spider, who sits moveless in the centre of the net. If the test is to be applied to a species with a daytime hiding-place amid the foliage, the dead Locust is laid on the web, more or less near the centre, no matter how.

In both cases, nothing happens at first. The Epeira remains in her motionless attitude, even when the morsel is at a short distance in front of her. She is indifferent to the presence of the game, does not seem to perceive it, so much so that she ends by wearing out my patience. Then, with a long straw, which enables me to conceal myself slightly, I set the dead insect trembling.

That is quite enough. The Banded Epeira and the Silky Epeira hasten to the central floor; the others come down from the branch; all go to the Locust, swathe him with tape, treat him, in short, as they would treat a live prey captured under normal conditions. It took the shaking of the web to decide them to attack.

Perhaps the grey colour of the Locust is not sufficiently conspicuous to attract attention by itself. Then let us try red, the brightest colour to our retina and probably also to the Spiders'. None of the game hunted by the Epeirae being clad in scarlet, I make a small bundle out of red wool, a bait of the size of a Locust. I glue it to the web.

My stratagem succeeds. As long as the parcel is stationary, the Spider is not roused; but, the moment it trembles, stirred by my straw, she runs up eagerly.

There are silly ones who just touch the thing with their legs and, without further enquiries, swathe it in silk after the manner of the usual game. They even go so far as to dig their fangs into the bait, following the rule of the preliminary poisoning. Then and then only the mistake is recognized and the tricked Spider retires and does not come back, unless it be long afterwards, when she flings the lumbersome object out of the web.

There are also clever ones. Like the others, these hasten to the red-woollen lure, which my straw insidiously keeps moving; they come from their tent among the leaves as readily as from the centre of the web; they explore it with their palpi and their legs; but, soon perceiving that the thing is valueless, they are careful not to spend their silk on useless bonds. My quivering bait does not deceive them. It is flung out after a brief inspection.

Still, the clever ones, like the silly ones, run even from a distance, from their leafy ambush. How do they know? Certainly not by sight. Before recognizing their mistake, they have to hold the object between their legs and even to nibble at it a little. They are extremely short-sighted. At a hand's-breadth's distance, the lifeless prey, unable to shake the web, remains unperceived. Besides, in many cases, the hunting takes place in the dense darkness of the night, when sight, even if it were good, would not avail.

If the eyes are insufficient guides, even close at hand, how will it be when the prey has to be spied from afar? In that case, an intelligence apparatus for long-distance work becomes indispensable. We have no difficulty in detecting the apparatus.

Let us look attentively behind the web of any Epeira with a daytime hiding-place: we shall see a thread that starts from the centre of the network, ascends in a slanting line outside the plane of the web and ends at the ambush where the Spider lurks all day. Except at the central point, there is no connection between this thread and the rest of the work, no interweaving with the scaffolding-threads. Free of impediment, the line runs straight from the centre of the net to the ambush-tent. Its length averages twenty-two inches. The Angular Epeira, settled high up in the trees, has shown me some as long as eight or nine feet.

There is no doubt that this slanting line is a foot-bridge which allows the Spider to repair hurriedly to the web, when summoned by urgent business, and then, when her round is finished, to return to her hut. In fact, it is the road which I see her follow, in going and coming. But is that all? No; for, if the Epeira had no aim in view but a means of rapid transit between her tent and the net, the foot-bridge would be fastened to the upper edge of the web. The journey would be shorter and the slope less steep.

Why, moreover, does this line always start in the centre of the sticky network and nowhere else? Because that is the point where the spokes meet and, therefore, the common centre of vibration. Anything that moves upon the web sets it shaking. All then that is needed is a thread issuing from this central point to convey to a distance the news of a prey struggling in some part or other of the net. The slanting cord, extending outside the plane of the web, is more than a foot-bridge: it is, above all, a signalling-apparatus, a telegraph-wire.

Let us try experiment. I place a Locust on the network. Caught in the sticky toils, he plunges about. Forthwith, the Spider issues impetuously from her hut, comes down the foot-bridge, makes a rush for the Locust, wraps him up and operates on him according to rule. Soon after, she hoists him, fastened by a line to her spinneret, and drags him to her hiding-place, where a long banquet will be held. So far, nothing new: things happen as usual.

I leave the Spider to mind her own affairs for some days before I interfere with her. I again propose to give her a Locust; but this time I first cut the signalling-thread with a touch of the scissors, without shaking any part of the edifice. The game is then laid on the web. Complete success: the entangled insect struggles, sets the net quivering; the Spider, on her side, does not stir, as though heedless of events.

The idea might occur to one that, in this business, the Epeira stays motionless in her cabin since she is prevented from hurrying down, because the foot-bridge is broken. Let us undeceive ourselves: for one road open to her there are a hundred, all ready to bring her to the place where her presence is now required. The network is fastened to the branches by a host of lines, all of them very easy to cross. Well, the Epeira embarks upon none of them, but remains moveless and self-absorbed.

Why? Because her telegraph, being out of order, no longer tells her of the shaking of the web. The captured prey is too far off for her to see it; she is all unwitting. A good hour passes, with the Locust still kicking, the Spider impassive, myself watching. Nevertheless, in the end, the Epeira wakes up: no longer feeling the signalling-thread, broken by my scissors, as taut as usual under her legs, she comes to look into the state of things. The web is reached, without the least difficulty, by one of the lines of the framework, the first that offers. The Locust is then perceived and forthwith enswathed, after which the signalling-thread is remade, taking the place of the one which I have broken. Along this road the Spider goes home, dragging her prey behind her.

My neighbour, the mighty Angular Epeira, with her telegraph-wire nine feet long, has even better things in store for me. One morning I find her web, which is now deserted, almost intact, a proof that the night's hunting has not been good. The animal must be hungry. With a piece of game for a bait, I hope to bring her down from her lofty retreat.

I entangle in the web a rare morsel, a Dragon-fly, who struggles desperately and sets the whole net a-shaking. The other, up above, leaves her lurking-place amid the cypress-foliage, strides swiftly down along her telegraph-wire, comes to the Dragon-fly, trusses her and at once climbs home again by the same road, with her prize dangling at her heels by a thread. The final sacrifice will take place in the quiet of the leafy sanctuary.

A few days later I renew my experiment under the same conditions, but, this time, I first cut the signalling-thread. In vain I select a large Dragon-fly, a very restless prisoner; in vain I exert my patience: the Spider does not come down all day. Her telegraph being broken, she receives no notice of what is happening nine feet below. The entangled morsel remains where it lies, not despised, but unknown. At nightfall the Epeira leaves her cabin, passes over the ruins of her web, finds the Dragon-fly and eats him on the spot, after which the net is renewed.

The Epeirae, who occupy a distant retreat by day, cannot do without a private wire that keeps them in permanent communication with the deserted web. All of them have one, in point of fact, but only when age comes, age prone to rest and to long slumbers. In their youth, the Epeirae, who are then very wide awake, know nothing of the art of telegraphy. Besides, their web, a short-lived work whereof hardly a trace remains on the morrow, does not allow of this kind of industry. It is no use going to the expense of a signalling-apparatus for a ruined snare wherein nothing can now be caught. Only the old Spiders, meditating or dozing in their green tent, are warned from afar, by telegraph, of what takes place on the web.

To save herself from keeping a close watch that would degenerate into drudgery and to remain alive to events even when resting, with her back turned on the net, the ambushed Spider always has her foot upon the telegraph-wire. Of my observations on this subject, let me relate the following, which will be sufficient for our purpose.

An Angular Epeira, with a remarkably fine belly, has spun her web between two laurustine-shrubs, covering a width of nearly a yard. The sun beats upon the snare, which is abandoned long before dawn. The Spider is in her day manor, a resort easily discovered by following the telegraph-wire. It is a vaulted chamber of dead leaves, joined together with a few bits of silk. The refuge is deep: the Spider disappears in it entirely, all but her rounded hind-quarters, which bar the entrance to her donjon.

With her front half plunged into the back of her hut, the Epeira certainly cannot see her web. Even if she had good sight, instead of being purblind, her position could not possibly allow her to keep the prey in view. Does she give up hunting during this period of bright sunlight? Not at all. Look again.

Wonderful! One of her hind-legs is stretched outside the leafy cabin; and the signalling-thread ends just at the tip of that leg. Whoso has not seen the Epeira in this attitude, with her hand, so to speak, on the telegraph-receiver, knows nothing of one of the most curious instances of animal cleverness. Let any game appear upon the scene; and the slumberer, forthwith aroused by means of the leg receiving the vibrations, hastens up. A Locust whom I myself lay on the web procures her this agreeable shock and what follows. If she is satisfied with her bag, I am still more satisfied with what I have learnt.

One word more. The web is often shaken by the wind. The different parts of the framework, tossed and teased by the eddying air-currents, cannot fail to transmit their vibration to the signalling-thread. Nevertheless, the Spider does not quit her hut and remains indifferent to the commotion prevailing in the net. Her line, therefore, is something better than a bell-rope that pulls and communicates the impulse given: it is a telephone capable, like our own, of transmitting infinitesimal waves of sound. Clutching her telephone-wire with a toe, the Spider listens with her leg; she perceives the innermost vibrations; she distinguishes between the vibration proceeding from a prisoner and the mere shaking caused by the wind.

CHAPTER 11.

THE EUMENES.

A wasp-like garb of motley black and yellow; a slender and graceful figure; wings not spread out flat, when resting, but folded lengthwise in two; the abdomen a sort of chemist's retort, which swells into a gourd and is fastened to the thorax by a long neck, first distending into a pear, then shrinking to a thread; a leisurely and silent flight; lonely habits. There we have a summary sketch of the Eumenes. My part of the country possesses two species: the larger, Eumenes Amedei, Lep., measures nearly an inch in length; the other, Eumenes pomiformis, Fabr., is a reduction of the first to the scale of one-half. (I include three species promiscuously under this one name, that is to say, Eumenes pomiformis, Fabr., E. bipunctis, Sauss., and E. dubius, Sauss. As I did not distinguish between them in my first investigations, which date a very long time back, it is not possible for me to ascribe to each of them its respective nest. But their habits are the same, for which reason this confusion does not injuriously affect the order of ideas in the present chapter.—Author's Note.)

Similar in form and colouring, both possess a like talent for architecture; and this talent is expressed in a work of the highest perfection which charms the most untutored eye. Their dwelling is a masterpiece. The Eumenes follow the profession of arms, which is unfavourable to artistic effort; they stab a prey with their sting; they pillage and plunder. They are predatory Hymenoptera, victualling their grubs with caterpillars. It will be interesting to compare their habits with those of the operator on the Grey Worm. (Ammophila hirsuta, who hunts the Grey Worm, the caterpillar of Noctua segetum, the Dart or Turnip Moth.—Translator's Note.)

Though the quarry—caterpillars in either case—remain the same, perhaps instinct, which is liable to vary with the species, has fresh glimpses in store for us. Besides, the edifice built by the Eumenes in itself deserves inspection.

The Hunting Wasps whose story we have described in former volumes are wonderfully well versed in the art of wielding the lancet; they astound us with their surgical methods, which they seem to have learnt from some physiologist who allows nothing to escape him; but those skilful slayers have no merit as builders of dwelling-houses. What is their home, in point of fact? An underground passage, with a cell at the end of it; a gallery, an excavation, a shapeless cave. It is miner's work, navvy's work: vigorous sometimes, artistic never. They use the pick-axe for loosening, the crowbar for shifting, the rake for extracting the materials, but never the trowel for laying. Now in the Eumenes we see real masons, who build their houses bit by bit with stone and mortar and run them up in the open, either on the firm rock or on the shaky support of a bough. Hunting alternates with architecture; the insect is a Nimrod or a Vitruvius by turns. (Marcus Vitruvius Pollio, the Roman architect and engineer.—Translator's Note.)

And, first of all, what sites do these builders select for their homes? Should you pass some little garden-wall, facing south, in a sun-scorched corner, look at the stones that are not covered with plaster, look at them one by one, especially the largest; examine the masses of boulders, at no great height from the ground, where the fierce rays have heated them to the temperature of a Turkish bath; and, perhaps, if you seek long enough, you will light upon the structure of Eumenes Amedei. The insect is scarce and lives apart; a meeting is an event upon which we must not count with too great confidence. It is an African species and loves the heat that ripens the carob and the date. It haunts the sunniest spots and selects rocks or firm stones as a foundation for its nest. Sometimes also, but seldom, it copies the Chalicodoma of the Walls and builds upon an ordinary pebble. (Or Mason-bee.—Translator's Note.)

Eumenes pomiformis is much more common and is comparatively indifferent to the nature of the foundation whereon she erects her cells. She builds on walls, on isolated stones, on the wood of the inner surface of half-closed shutters; or else she adopts an aerial base, the slender twig of a shrub, the withered sprig of a plant of some sort. Any form of support serves her purpose. Nor does she trouble about shelter. Less chilly than her African cousin, she does not shun the unprotected spaces exposed to every wind that blows.

When erected on a horizontal surface, where nothing interferes with it, the structure of Eumenes Amedei is a symmetrical cupola, a spherical skull-cap, with, at

the top, a narrow passage just wide enough for the insect, and surmounted by a
neatly funnelled neck. It suggests the round hut of the Eskimo or of the ancient
Gael, with its central chimney. Two centimetres and a half (.97 inch.—
Translator's Note.), more or less, represent the diameter, and two centimetres the
height. (.78 inch.—Translator's Note.) When the support is a perpendicular
plane, the building still retains the domed shape, but the entrance- and exit-fun-
nel opens at the side, upwards. The floor of this apartment calls for no labour: it
is supplied direct by the bare stone.

Having chosen the site, the builder erects a circular fence about three millimetres
thick. (.118 inch.—Translator's Note.) The materials consist of mortar and small
stones. The insect selects its stone-quarry in some well-trodden path, on some
neighbouring road, at the driest, hardest spots. With its mandibles, it scrapes
together a small quantity of dust and saturates it with saliva until the whole
becomes a regular hydraulic mortar which soon sets and is no longer susceptible
to water. The Mason-bees have shown us a similar exploitation of the beaten paths
and of the road-mender's macadam. All these open-air builders, all these erectors
of monuments exposed to wind and weather require an exceedingly dry stone-
dust; otherwise the material, already moistened with water, would not properly
absorb the liquid that is to give it cohesion; and the edifice would soon be
wrecked by the rains. They possess the sense of discrimination of the plasterer,
who rejects plaster injured by damp. We shall see presently how the insects that
build under shelter avoid this laborious macadam-scraping and give the preference to fresh earth already reduced to a paste by its own dampness. When common lime answers our purpose, we do not trouble about Roman cement. Now
Eumenes Amedei requires a first-class cement, even better than that of the
Chalicodoma of the Walls, for the work, when finished, does not receive the thick
covering wherewith the Mason-bee protects her cluster of cells. And therefore the
cupola-builder, as often as she can, uses the highway as her stone-pit.

With the mortar, flints are needed. These are bits of gravel of an almost unvary-
ing size—that of a peppercorn—but of a shape and kind differing greatly, according to the places worked. Some are sharp-cornered, with facets determined by
chance fractures; some are round, polished by friction under water. Some are of
limestone, others of silicic matter. The favourite stones, when the neighbourhood
of the nest permits, are little nodules of quartz, smooth and semitransparent.
These are selected with minute care. The insect weighs them, so to say, measures
them with the compass of its mandibles and does not accept them until after recognizing in them the requisite qualities of size and hardness.

A circular fence, we were saying, is begun on the bare rock. Before the mortar sets,
which does not take long, the mason sticks a few stones into the soft mass, as the

work advances. She dabs them half-way into the cement, so as to leave them jutting out to a large extent, without penetrating to the inside, where the wall must remain smooth for the sake of the larva's comfort. If necessary, a little plaster is added, to tone down the inner protuberances. The solidly embedded stonework alternates with the pure mortarwork, of which each fresh course receives its facing of tiny encrusted pebbles. As the edifice is raised, the builder slopes the construction a little towards the centre and fashions the curve which will give the spherical shape. We employ arched centrings to support the masonry of a dome while building: the Eumenes, more daring than we, erects her cupola without any scaffolding.

A round orifice is contrived at the summit; and, on this orifice, rises a funnelled mouthpiece built of pure cement. It might be the graceful neck of some Etruscan vase. When the cell is victualled and the egg laid, this mouthpiece is closed with a cement plug; and in this plug is set a little pebble, one alone, no more: the ritual never varies. This work of rustic architecture has naught to fear from the inclemency of the weather; it does not yield to the pressure of the fingers; it resists the knife that attempts to remove it without breaking it. Its nipple shape and the bits of gravel wherewith it bristles all over the outside remind one of certain cromlechs of olden time, of certain tumuli whose domes are strewn with Cyclopean stones.

Such is the appearance of the edifice when the cell stands alone; but the Hymenopteron nearly always fixes other domes against her first, to the number of five, six, or more. This shortens the labour by allowing her to use the same partition for two adjoining rooms. The original elegant symmetry is lost and the whole now forms a cluster which, at first sight, appears to be merely a clod of dry mud, sprinkled with tiny pebbles. But let us examine the shapeless mass more closely and we shall perceive the number of chambers composing the habitation with the funnelled mouths, each quite distinct and each furnished with its gravel stopper set in the cement.

The Chalicodoma of the Walls employs the same building methods as Eumenes Amedei: in the courses of cement she fixes, on the outside, small stones of minor bulk. Her work begins by being a turret of rustic art, not without a certain prettiness; then, when the cells are placed side by side, the whole construction degenerates into a lump governed apparently by no architectural rule. Moreover, the Mason-bee covers her mass of cells with a thick layer of cement, which conceals the original rockwork edifice. The Eumenes does not resort to this general coating: her building is too strong to need it; she leaves the pebbly facings uncovered, as well as the entrances to the cells. The two sorts of nests, although constructed of similar materials, are therefore easily distinguished.

The Eumenes' cupola is the work of an artist; and the artist would be sorry to cover his masterpiece with whitewash. I crave forgiveness for a suggestion which I advance with all the reserve befitting so delicate a subject. Would it not be possible for the cromlech-builder to take a pride in her work, to look upon it with some affection and to feel gratified by this evidence of her cleverness? Might there not be an insect science of aesthetics? I seem at least to catch a glimpse, in the Eumenes, of a propensity to beautify her work. The nest must be, before all, a solid habitation, an inviolable stronghold; but, should ornament intervene without jeopardizing the power of resistance, will the worker remain indifferent to it? Who would say?

Let us set forth the facts. The orifice at the top, if left as a mere hole, would suit the purpose quite as well as an elaborate door: the insect would lose nothing in regard to facilities for coming and going and would gain by shortening the labour. Yet we find, on the contrary, the mouth of an amphora, gracefully curved, worthy of a potter's wheel. A choice cement and careful work are necessary for the confection of its slender, funnelled shaft. Why this nice finish, if the builder be wholly absorbed in the solidity of her work?

Here is another detail: among the bits of gravel employed for the outer covering of the cupola, grains of quartz predominate. They are polished and translucent; they glitter slightly and please the eye. Why are these little pebbles preferred to chips of lime-stone, when both materials are found in equal abundance around the nest?

A yet more remarkable feature: we find pretty often, encrusted on the dome, a few tiny, empty snail-shells, bleached by the sun. The species usually selected by the Eumenes is one of the smaller Helices—Helix strigata—frequent on our parched slopes. I have seen nests where this Helix took the place of pebbles almost entirely. They were like boxes made of shells, the work of a patient hand.

A comparison offers here. Certain Australian birds, notably the Bower-birds, build themselves covered walks, or playhouses, with interwoven twigs, and decorate the two entrances to the portico by strewing the threshold with anything that they can find in the shape of glittering, polished, or bright-coloured objects. Every door-sill is a cabinet of curiosities where the collector gathers smooth pebbles, variegated shells, empty snail-shells, parrot's feathers, bones that have come to look like sticks of ivory. The odds and ends mislaid by man find a home in the bird's museum, where we see pipe-stems, metal buttons, strips of cotton stuff and stone axe-heads.

The collection at either entrance to the bower is large enough to fill half a bushel. As these objects are of no use to the bird, its only motive for accumulating them must be an art-lover's hobby. Our common Magpie has similar tastes: any shiny thing that he comes upon he picks up, hides and hoards.

Well, the Eumenes, who shares this passion for bright pebbles and empty snail-shells, is the Bower-bird of the insect world; but she is a more practical collector, knows how to combine the useful and the ornamental and employs her finds in the construction of her nest, which is both a fortress and a museum. When she finds nodules of translucent quartz, she rejects everything else: the building will be all the prettier for them. When she comes across a little white shell, she hastens to beautify her dome with it; should fortune smile and empty snail-shells abound, she encrusts the whole fabric with them, until it becomes the supreme expression of her artistic taste. Is this so? Or is it not so? Who shall decide?

The nest of Eumenes pomiformis is the size of an average cherry and constructed of pure mortar, without the least outward pebblework. Its shape is exactly similar to that which we have just described. When built upon a horizontal base of sufficient extent, it is a dome with a central neck, funnelled like the mouth of an urn. But when the foundation is reduced to a mere point, as on the twig of a shrub, the nest becomes a spherical capsule, always, of course, surmounted by a neck. It is then a miniature specimen of exotic pottery, a paunchy alcarraza. Its thickens is very slight, less than that of a sheet of paper; it crushes under the least effort of the fingers. The outside is not quite even. It displays wrinkles and seams, due to the different courses of mortar, or else knotty protuberances distributed almost concentrically.

Both Hymenoptera accumulate caterpillars in their coffers, whether domes or jars. Let us give an abstract of the bill of fare. These documents, for all their dryness, possess a value; they will enable whoso cares to interest himself in the Eumenes to perceive to what extent instinct varies the diet, according to the place and season. The food is plentiful, but lacks variety. It consists of tiny caterpillars, by which I mean the grubs of small Butterflies. We learn this from the structure, for we observe in the prey selected by either Hymenopteran the usual caterpillar organism. The body is composed of twelve segments, not including the head. The first three have true legs, the next two are legless, then come two segments with prolegs, two legless segments and, lastly, a terminal segment with prolegs. It is exactly the same structure which we saw in the Ammophila's Grey Worm.

My old notes give the following description of the caterpillars found in the nest of Eumenes Amedei: "a pale green or, less often, a yellowish body, covered with short white hairs; head wider than the front segment, dead-black and also bristling with hairs. Length: 16 to 18 millimetres (.63 to .7 inch.—Translator's Note.); width: about 3 millimetres." (.12 inch.—Translator's Note.) A quarter of a century and more has elapsed since I jotted down this descriptive sketch; and to-day, at Sérignan, I find in the Eumenes' larder the same game which I noticed long ago at Carpentras. Time and distance have not altered the nature of the provisions.

The number of morsels served for the meal of each larva interests us more than the quality. In the cells of Eumenes Amedei, I find sometimes five caterpillars and sometimes ten, which means a difference of a hundred per cent in the quantity of the food, for the morsels are of exactly the same size in both cases. Why this unequal supply, which gives a double portion to one larva and a single portion to another? The diners have the same appetite: what one nurseling demands a second must demand, unless we have here a different menu, according to the sexes. In the perfect stage the males are smaller than the females, are hardly half as much in weight or volume. The amount of victuals, therefore, required to bring them to their final development may be reduced by one-half. In that case, the well-stocked cells belong to females; the others, more meagrely supplied, belong to males.

But the egg is laid when the provisions are stored; and this egg has a determined sex, though the most minute examination is not able to discover the differences which will decide the hatching of a female or a male. We are therefore needs driven to this strange conclusion: the mother knows beforehand the sex of the egg which she is about to lay; and this knowledge allows her to fill the larder according to the appetite of the future grub. What a strange world, so wholly different from ours! We fall back upon a special sense to explain the Ammophila's hunting; what can we fall back upon to account for this intuition of the future? Can the theory of chances play a part in the hazy problem? If nothing is logically arranged with a foreseen object, how is this clear vision of the invisible acquired?

The capsules of Eumenes pomiformis are literally crammed with game. It is true that the morsels are very small. My notes speak of fourteen green caterpillars in one cell and sixteen in a second cell. I have no other information about the integral diet of this Wasp, whom I have neglected somewhat, preferring to study her cousin, the builder of rockwork domes. As the two sexes differ in size, although to a lesser degree than in the case of Eumenes Amedei, I am inclined to think that those two well-filled cells belonged to females and that the males' cells must have a less sumptuous table. Not having seen for myself, I am content to set down this mere suspicion.

What I have seen and often seen is the pebbly nest, with the larva inside and the provisions partly consumed. To continue the rearing at home and follow my charge's progress from day to day was a business which I could not resist; besides, as far as I was able to see, it was easily managed. I had had some practice in this foster-father's trade; my association with the Bembex, the Ammophila, the Sphex (three species of Digger-wasps.—Translator's Note.) and many others had turned me into a passable insect-rearer. I was no novice in the art of dividing an old pen-box into compartments in which I laid a bed of sand and, on this bed, the larva and her provisions delicately removed from the maternal cell. Success was almost certain at each attempt: I used to watch the larvae at their meals, I saw my nurselings grow up and spin their cocoons. Relying upon the experience thus gained, I reckoned upon success in raising my Eumenes.

The results, however, in no way answered to my expectations. All my endeavours failed; and the larva allowed itself to die a piteous death without touching its provisions.

I ascribed my reverse to this, that and the other cause: perhaps I had injured the frail grub when demolishing the fortress; a splinter of masonry had bruised it when I forced open the hard dome with my knife; a too sudden exposure to the sun had surprised it when I withdrew it from the darkness of its cell; the open air might have dried up its moisture. I did the best I could to remedy all these probable reasons of failure. I went to work with every possible caution in breaking open the home; I cast the shadow of my body over the nest, to save the grub from sunstroke; I at once transferred larva and provisions into a glass tube and placed this tube in a box which I carried in my hand, to minimize the jolting on the journey. Nothing was of avail: the larva, when taken from its dwelling, always allowed itself to pine away.

For a long time I persisted in explaining my want of success by the difficulties attending the removal. Eumenes Amedei's cell is a strong casket which cannot be forced without sustaining a shock; and the demolition of a work of this kind entails such varied accidents that we are always liable to think that the worm has been bruised by the wreckage. As for carrying home the nest intact on its support, with a view to opening it with greater care than is permitted by a rough-and-ready operation in the fields, that is out of the question: the nest nearly always stands on an immovable rock or on some big stone forming part of a wall. If I failed in my attempts at rearing, it was because the larva had suffered when I was breaking up her house. The reason seemed a good one; and I let it go at that.

In the end, another idea occurred to me and made me doubt whether my rebuffs were always due to clumsy accidents. The Eumenes' cells are crammed with game:

there are ten caterpillars in the cell of Eumenes Amedei and fifteen in that of Eumenes pomiformis. These caterpillars, stabbed no doubt, but in a manner unknown to me, are not entirely motionless. The mandibles seize upon what is presented to them, the body buckles and unbuckles, the hinder half lashes out briskly when stirred with the point of a needle. At what spot is the egg laid amid that swarming mass, where thirty mandibles can make a hole in it, where a hundred and twenty pairs of legs can tear it? When the victuals consist of a single head of game, these perils do not exist; and the egg is laid on the victim not at hazard, but upon a judiciously chosen spot. Thus, for instance, Ammophila hirsuta fixes hers, by one end, cross-wise, on the Grey Worm, on the side of the first prolegged segment. The eggs hang over the caterpillar's back, away from the legs, whose proximity might be dangerous. The worm, moreover, stung in the greater number of its nerve-centres, lies on one side, motionless and incapable of bodily contortions or said an jerks of its hinder segments. If the mandibles try to snap, if the legs give a kick or two, they find nothing in front of them: the Ammophila's egg is at the opposite side. The tiny grub is thus able, as soon as it hatches, to dig into the giant's belly in full security.

How different are the conditions in the Eumenes' cell. The caterpillars are imperfectly paralysed, perhaps because they have received but a single stab; they toss about when touched with a pin; they are bound to wriggle when bitten by the larva. If the egg is laid on one of them, the first morsel will, I admit, be consumed without danger, on condition that the point of attack be wisely chosen; but there remain others which are not deprived of every means of defence. Let a movement take place in the mass; and the egg, shifted from the upper layer, will tumble into a pitfall of legs and mandibles. The least thing is enough to jeopardize its existence; and this least thing has every chance of being brought about in the disordered heap of caterpillars. The egg, a tiny cylinder, transparent as crystal, is extremely delicate: a touch withers it, the least pressure crushes it.

No, its place is not in the mass of provisions, for the caterpillars, I repeat, are not sufficiently harmless. Their paralysis is incomplete, as is proved by their contortions when I irritate them and shown, on the other hand, by a very important fact. I have sometimes taken from Eumenes Amedei's cell a few heads of game half transformed into chrysalids. It is evident that the transformation was effected in the cell itself and, therefore, after the operation which the Wasp had performed upon them. Whereof does this operation consist? I cannot say precisely, never having seen the huntress at work. The sting most certainly has played its part; but where? And how often? This is what we do not know. What we are able to declare is that the torpor is not very deep, inasmuch as the patient sometimes retains enough vitality to shed its skin and become a chrysalid. Everything thus tends to make us ask by what stratagem the egg is shielded from danger.

This stratagem I longed to discover; I would not be put off by the scarcity of nests, by the irksomeness of the searches, by the risk of sunstroke, by the time taken up, by the vain breaking open of unsuitable cells; I meant to see and I saw. Here is my method: with the point of a knife and a pair of nippers, I make a side opening, a window, beneath the dome of Eumenes Amedei and Eumenes pomiformis. I work with the greatest care, so as not to injure the recluse. Formerly I attacked the cupola from the top, now I attack it from the side. I stop when the breach is large enough to allow me to see the state of things within.

What is this state of things? I pause to give the reader time to reflect and to think out for himself a means of safety that will protect the egg and afterwards the grub in the perilous conditions which I have set forth. Seek, think and contrive, such of you as have inventive minds. Have you guessed it? Do you give it up? I may as well tell you.

The egg is not laid upon the provisions; it is hung from the top of the cupola by a thread which vies with that of a Spider's web for slenderness. The dainty cylinder quivers and swings to and fro at the least breath; it reminds me of the famous pendulum suspended from the dome of the Pantheon to prove the rotation of the earth. The victuals are heaped up underneath.

Second act of this wondrous spectacle. In order to witness it, we must open a window in cell upon cell until fortune deigns to smile upon us. The larva is hatched and already fairly large. Like the egg, it hangs perpendicularly, by the rear, from the ceiling; but the suspensory cord has gained considerably in length and consists of the original thread eked out by a sort of ribbon. The grub is at dinner: head downwards, it is digging into the limp belly of one of the caterpillars. I touch up the game that is still intact with a straw. The caterpillars grow restless. The grub forthwith retires from the fray. And how? Marvel is added to marvels: what I took for a flat cord, for a ribbon, at the lower end of the suspensory thread, is a sheath, a scabbard, a sort of ascending gallery wherein the larva crawls backwards and makes its way up. The cast shell of the egg, retaining its cylindrical form and perhaps lengthened by a special operation on the part of the new-born grub, forms this safety-channel. At the least sign of danger in the heap of caterpillars, the larva retreats into its sheath and climbs back to the ceiling, where the swarming rabble cannot reach it. When peace is restored, it slides down its case and returns to table, with its head over the viands and its rear upturned and ready to withdraw in case of need.

Third and last act. Strength has come; the larva is brawny enough not to dread the movements of the caterpillars' bodies. Besides, the caterpillars, mortified by fasting and weakened by a prolonged torpor, become more and more incapable of defence. The perils of the tender babe are succeeded by the security of the lusty stripling; and the grub, henceforth scorning its sheathed lift, lets itself drop upon the game that remains. And thus the banquet ends in normal fashion.

That is what I saw in the nests of both species of the Eumenes and that is what I showed to friends who were even more surprised than I by these ingenious tactics. The egg hanging from the ceiling, at a distance from the provisions, has naught to fear from the caterpillars, which flounder about below. The new-hatched larva, whose suspensory cord is lengthened by the sheath of the egg, reaches the game and takes a first cautious bite at it. If there be danger, it climbs back to the ceiling by retreating inside the scabbard. This explains the failure of my earlier attempts. Not knowing of the safety-thread, so slender and so easily broken, I gathered at one time the egg, at another the young larva, after my inroads at the top had caused them to fall into the middle of the live victuals. Neither of them was able to thrive when brought into direct contact with the dangerous game.

If any one of my readers, to whom I appealed just now, has thought out something better than the Eumenes' invention, I beg that he will let me know: there is a curious parallel to be drawn between the inspirations of reason and the inspirations of instinct.

CHAPTER 12.

THE OSMIAE.

THEIR HABITS.

February has its sunny days, heralding spring, to which rude winter will reluctantly yield place. In snug corners, among the rocks, the great spurge of our district, the characias of the Greeks, the jusclo of the Provençals, begins to lift its drooping inflorescence and discreetly opens a few sombre flowers. Here the first midges of the year will come to slake their thirst. By the time that the tip of the stalks reaches the perpendicular, the worst of the cold weather will be over.

Another eager one, the almond-tree, risking the loss of its fruit, hastens to echo these preludes to the festival of the sun, preludes which are too often treacherous. A few days of soft skies and it becomes a glorious dome of white flowers, each twinkling with a roseate eye. The country, which still lacks green, seems dotted everywhere with white-satin pavilions. 'Twould be a callous heart indeed that could resist the magic of this awakening.

The insect nation is represented at these rites by a few of its more zealous members. There is first of all the Honey-bee, the sworn enemy of strikes, who profits by the least lull of winter to find out if some rosemary or other is not beginning to open somewhere near the hive. The droning of the busy swarms fills the flowery vault, while a snow of petals falls softly to the foot of the tree.

Together with the population of harvesters there mingles another, less numerous, of mere drinkers, whose nesting-time has not yet begun. This is the colony of the

Osmiae, those exceedingly pretty solitary bees, with their copper-coloured skin and bright-red fleece. Two species have come hurrying up to take part in the joys of the almond-tree: first, the Horned Osmia, clad in black velvet on the head and breast, with red velvet on the abdomen; and, a little later, the Three-horned Osmia, whose livery must be red and red only. These are the first delegates despatched by the pollen-gleaners to ascertain the state of the season and attend the festival of the early blooms.

'Tis but a moment since they burst their cocoon, the winter abode: they have left their retreats in the crevices of the old walls; should the north wind blow and set the almond-tree shivering, they will hasten to return to them. Hail to you, O my dear Osmiae, who yearly, from the far end of the harmas, opposite snow-capped Ventoux (A mountain in the Provençal Alps, near Carpentras and Sérignan 6,271 feet.—Translator's Note.), bring me the first tidings of the awakening of the insect world! I am one of your friends; let us talk about you a little.

Most of the Osmiae of my region do not themselves prepare the dwelling destined for the laying. They want ready-made lodgings, such as the old cells and old galleries of Anthophorae and Chalicodomae. If these favourite haunts are lacking, then a hiding-place in the wall, a round hole in some bit of wood, the tube of a reed, the spiral of a dead Snail under a heap of stones are adopted, according to the tastes of the several species. The retreat selected is divided into chambers by partition-walls, after which the entrance to the dwelling receives a massive seal. That is the sum-total of the building done.

For this plasterer's rather than mason's work, the Horned and the Three-horned Osmia employ soft earth. This material is a sort of dried mud, which turns to pap on the addition of a drop of water. The two Osmiae limit themselves to gathering natural soaked earth, mud in short, which they allow to dry without any special preparation on their part; and so they need deep and well-sheltered retreats, into which the rain cannot penetrate, or the work would fall to pieces.

Latreille's Osmia uses different materials for her partitions and her doors. She chews the leaves of some mucilaginous plant, some mallow perhaps, and then prepares a sort of green putty with which she builds her partitions and finally closes the entrance to the dwelling. When she settles in the spacious cells of the Masked Anthophora (Anthophora personata, Illig.), the entrance to the gallery, which is wide enough to admit a man's finger, is closed with a voluminous plug of this vegetable paste. On the earthy banks, hardened by the sun, the home is then betrayed by the gaudy colour of the lid. It is as though the authorities had closed the door and affixed to it their great seals of green wax.

So far then as their building-materials are concerned, the Osmiae whom I have been able to observe are divided into two classes: one building compartments with mud, the other with a green-tinted vegetable putty. To the latter belongs Latreille's Osmia. The first section includes the Horned Osmia and the Three-horned Osmia, both so remarkable for the horny tubercles on their faces.

The great reed of the south, Arundo donax, is often used, in the country, for making rough garden-shelters against the mistral or just for fences. These reeds, the ends of which are chopped off to make them all the same length, are planted perpendicularly in the earth. I have often explored them in the hope of finding Osmia-nests. My search has very seldom succeeded. The failure is easily explained. The partitions and the closing-plug of the Horned and of the Three-horned Osmia are made, as we have seen, of a sort of mud which water instantly reduces to pap. With the upright position of the reeds, the stopper of the opening would receive the rain and would become diluted; the ceilings of the storeys would fall in and the family would perish by drowning. Therefore the Osmia, who knew of these drawbacks before I did, refuses the reeds when they are placed perpendicularly.

The same reed is used for a second purpose. We make canisses of it, that is to say, hurdles, which, in spring, serve for the rearing of Silkworms and, in autumn, for the drying of figs. At the end of April and during May, which is the time when the Osmiae work, the canisses are indoors, in the Silkworm nurseries, where the Bee cannot take possession of them; in autumn, they are outside, exposing their layers of figs and peeled peaches to the sun; but by that time the Osmiae have long disappeared. If, however, during the spring, an old, disused hurdle is left out of doors, in a horizontal position, the Three-horned Osmia often takes possession of it and makes use of the two ends, where the reeds lie truncated and open.

There are other quarters that suit the Three-horned Osmia, who is not particular, it seems to me, and will make shift with any hiding-place, so long as it have the requisite conditions of diameter, solidity, sanitation and kindly darkness. The most original dwellings that I know her to occupy are disused Snail-shells, especially the house of the Common Snail (Helix aspersa). Let us go to the slope of the hills thick with olive-trees and inspect the little supporting-walls which are built of dry stones and face the south. In the crevices of this insecure masonry we shall reap a harvest of old Snail-shells, plugged with earth right up to the orifice. The family of the Three-horned Osmia is settled in the spiral of those shells, which is subdivided into chambers by mud partitions.

The Three-pronged Osmia (O. Tridentata, Duf. and Per.) alone creates a home of her own, digging herself a channel with her mandibles in dry bramble and sometimes in danewort.

The Osmia loves mystery. She wants a dark retreat, hidden from the eye. I would like, nevertheless, to watch her in the privacy of her home and to witness her work with the same facility as if she were nest-building in the open air. Perhaps there are some interesting characteristics to be picked up in the depths of her retreats. It remains to be seen whether my wish can be realized.

When studying the insect's mental capacity, especially its very retentive memory for places, I was led to ask myself whether it would not be possible to make a suitably-chosen Bee build in any place that I wished, even in my study. And I wanted, for an experiment of this sort, not an individual but a numerous colony. My preference lent towards the Three-horned Osmia, who is very plentiful in my neighbourhood, where, together with Latreille's Osmia, she frequents in particular the monstrous nests of the Chalicodoma of the Sheds. I therefore thought out a scheme for making the Three-horned Osmia accept my study as her settlement and build her nest in glass tubes, through which I could easily watch the progress. To these crystal galleries, which might well inspire a certain distrust, were to be added more natural retreats: reeds of every length and thickness and disused Chalicodoma-nests taken from among the biggest and the smallest. A scheme like this sounds mad. I admit it, while mentioning that perhaps none ever succeeded so well with me. We shall see as much presently.

My method is extremely simple. All I ask is that the birth of my insects, that is to say, their first seeing the light, their emerging from the cocoon, should take place on the spot where I propose to make them settle. Here there must be retreats of no matter what nature, but of a shape similar to that in which the Osmia delights. The first impressions of sight, which are the most long-lived of any, shall bring back my insects to the place of their birth. And not only will the Osmiae return, through the always open windows, but they will also nidify on the natal spot, if they find something like the necessary conditions.

And so, all through the winter, I collect Osmia-cocoons picked up in the nests of the Mason-bee of the Sheds; I go to Carpentras to glean a more plentiful supply in the nests of the Anthophora. I spread out my stock in a large open box on a table which receives a bright diffused light but not the direct rays of the sun. The table stands between two windows facing south and overlooking the garden. When the moment of hatching comes, those two windows will always remain open to give the swarm entire liberty to go in and out as it pleases. The glass tubes and reed-stumps are laid here and there, in fine disorder, close to the heaps of cocoons and all in a horizontal position, for the Osmia will have nothing to do with upright reeds. Although such a precaution is not indispensable, I take care to place some cocoons in each cylinder. The hatching of some of the Osmiae will therefore take place under cover of the galleries destined to be the building-yard later; and the site will be all the more deeply impressed on their memory. When I have made these comprehensive arrangements, there is nothing more to be done; and I wait patiently for the building-season to open.

My Osmiae leave their cocoons in the second half of April. Under the immediate rays of the sun, in well-sheltered nooks, the hatching would occur a month earlier, as we can see from the mixed population of the snowy almond-tree. The constant shade in my study has delayed the awakening, without, however, making any change in the nesting-period, which synchronizes with the flowering of the thyme. We now have, around my working-table, my books, my jars and my various appliances, a buzzing crowd that goes in and out of the windows at every moment. I enjoin the household henceforth not to touch a thing in the insects' laboratory, to do no more sweeping, no more dusting. They might disturb a swarm and make it think that my hospitality was not to be trusted. During four or five weeks I witness the work of a number of Osmiae which is much too large to allow my watching their individual operations. I content myself with a few, whom I mark with different-coloured spots to distinguish them; and I take no notice of the others, whose finished work will have my attention later.

The first to appear are the males. If the sun is bright, they flutter around the heap of tubes as if to take careful note of the locality; blows are exchanged and the rival swains indulge in mild skirmishing on the floor, then shake the dust off their wings. They fly assiduously from tube to tube, placing their heads in the orifices to see if some female will at last make up her mind to emerge.

One does, in point of fact. She is covered with dust and has the disordered toilet that is inseparable from the hard work of the deliverance. A lover has seen her, so has a second, likewise a third. All crowd round her. The lady responds to their advances by clashing her mandibles, which open and shut rapidly, several times in succession. The suitors forthwith fall back; and they also, no doubt to keep up their dignity, execute savage mandibular grimaces. Then the beauty retires into the arbour and her wooers resume their places on the threshold. A fresh appearance of the female, who repeats the play with her jaws; a fresh retreat of the males, who do the best they can to flourish their own pincers. The Osmiae have a strange way of declaring their passion: with that fearsome gnashing of their mandibles, the lovers look as though they meant to devour each other. It suggests the thumps affected by our yokels in their moments of gallantry.

The ingenuous idyll is soon over. The females, who grow more numerous from day to day, inspect the premises; they buzz outside the glass galleries and the reed dwellings; they go in, stay for a while, come out, go in again and then fly away briskly into the garden. They return, first one, then another. They halt outside, in the sun, or on the shutters fastened back against the wall; they hover in the window-recess, come inside, go to the reeds and give a glance at them, only to set off again and to return soon after. Thus do they learn to know their home, thus do they fix their birthplace in their memory. The village of our childhood is always a cherished spot, never to be effaced from our recollection. The Osmia's life endures for a month; and she acquires a lasting remembrance of her hamlet in a couple of days. 'Twas there that she was born; 'twas there that she loved; 'tis there that she will return. Dulces reminiscitur Argos.

> (Now falling by another's wound, his eyes
> He casts to heaven, on Argos thinks and dies.
> —"Aeneid" Book 10, Dryden's translation.)

At last each has made her choice. The work of construction begins; and my expectations are fulfilled far beyond my wishes. The Osmiae build nests in all the retreats which I have placed at their disposal. And now, O my Osmiae, I leave you a free field!

The work begins with a thorough spring-cleaning of the home. Remnants of cocoons, dirt consisting of spoilt honey, bits of plaster from broken partitions, remains of dried Mollusc at the bottom of a shell: these and much other insanitary refuse must first of all disappear. Violently the Osmia tugs at the offending object and tears it out; and then off she goes in a desperate hurry, to dispose of it far away from the study. They are all alike, these ardent sweepers: in their excessive zeal, they fear lest they should block up the speck of dust which they might drop in front of the new house. The glass tubes, which I myself have rinsed under the tap, are not exempt from a scrupulous cleaning. The Osmia dusts them, brushes them thoroughly with her tarsi and then sweeps them out backwards. What does she pick up? Not a thing. It makes no difference: as a conscientious housewife, she gives the place a touch of the broom nevertheless.

Now for the provisions and the partition-walls. Here the order of the work changes according to the diameter of the cylinder. My glass tubes vary greatly in dimensions. The largest have an inner width of a dozen millimetres (Nearly half an inch.—Translator's Note.); the narrowest measure six or seven. (About a quarter of an inch.—Translator's Note.) In the latter, if the bottom suit her, the Osmia sets to work bringing pollen and honey. If the bottom do not suit her, if the sorghum-pith plug with which I have closed the rear-end of the tube be too irregular and badly-joined, the Bee coats it with a little mortar. When this small repair is made, the harvesting begins.

In the wider tubes, the work proceeds quite differently. At the moment when the Osmia disgorges her honey and especially at the moment when, with her hind-tarsi, she rubs the pollen-dust from her ventral brush, she needs a narrow aperture, just big enough to allow of her passage. I imagine that in a straitened gallery the rubbing of her whole body against the sides gives the harvester a support for her brushing-work. In a spacious cylinder this support fails her; and the Osmia starts with creating one for herself, which she does by narrowing the channel. Whether it be to facilitate the storing of the victuals or for any other reason, the fact remains that the Osmia housed in a wide tube begins with the partitioning.

Her division is made by a dab of clay placed at right angles to the axis of the cylinder, at a distance from the bottom determined by the ordinary length of a cell. The wad is not a complete round; it is more crescent-shaped, leaving a circular space between it and one side of the tube. Fresh layers are swiftly added to the dab of clay; and soon the tube is divided by a partition which has a circular opening at the side of it, a sort of dog-hole through which the Osmia will proceed to knead the Bee-bread. When the victualling is finished and the egg laid upon the heap, the whole is closed and the filled-up partition becomes the bottom of the next cell. Then the same method is repeated, that is to say, in front of the just completed ceiling a second partition is built, again with a side-passage, which is stouter, owing to its distance from the centre, and better able to withstand the numerous comings and goings of the housewife than a central orifice, deprived of the direct support of the wall, could hope to be. When this partition is ready, the provisioning of the second cell is effected; and so on until the wide cylinder is completely stocked.

The building of this preliminary party-wall, with a narrow, round dog-hole, for a chamber to which the victuals will not be brought until later is not restricted to the Three-horned Osmia; it is also frequently found in the case of the Horned Osmia and of Latreille's Osmia. Nothing could be prettier than the work of the last-named, who goes to the plants for her material and fashions a delicate sheet in which she cuts a graceful arch. The Chinaman partitions his house with paper screens; Latreille's Osmia divides hers with disks of thin green cardboard perforated with a serving-hatch which remains until the room is completely furnished. When we have no glass houses at our disposal, we can see these little architectural refinements in the reeds of the hurdles, if we open them at the right season.

By splitting the bramble-stumps in the course of July, we perceive also that the Three-pronged Osmia notwithstanding her narrow gallery, follows the same practice as Latreille's Osmia, with a difference. She does not build a party-wall, which the diameter of the cylinder would not permit; she confines herself to putting up a frail circular pad of green putty, as though to limit, before any attempt at harvesting, the space to be occupied by the Bee-bread, whose depth could not be calculated afterwards if the insect did not first mark out its confines.

If, in order to see the Osmia's nest as a whole, we split a reed lengthwise, taking care not to disturb its contents; or, better still, if we select for examination the string of cells built in a glass tube, we are forthwith struck by one detail, namely, the uneven distances between the partitions, which are placed almost at right angles to the axis of the cylinder. It is these distances which fix the size of the chambers, which, with a similar base, have different heights and consequently unequal holding-capacities. The bottom partitions, the oldest, are farther apart; those of the front part, near the orifice, are closer together. Moreover, the provisions are plentiful in the loftier cells, whereas they are niggardly and reduced to one-half or even one-third in the cells of lesser height. Let me say at once that the large cells are destined for the females and the small ones for the males.

DISTRIBUTION OF THE SEXES.

Does the insect which stores up provisions proportionate to the needs of the egg which it is about to lay know beforehand the sex of that egg? Or is the truth even more paradoxical? What we have to do is to turn this suspicion into a certainty demonstrated by experiment. And first let us find out how the sexes are arranged.

It is not possible to ascertain the chronological order of a laying, except by going to suitably-chosen species. Fortunately there are a few species in which we do not find this difficulty: these are the Bees who keep to one gallery and build their cells in storeys. Among the number are the different inhabitants of the bramble-stumps, notably the Three-pronged Osmiae, who form an excellent subject for observation, partly because they are of imposing size—bigger than any other bramble-dwellers in my neighbourhood—partly because they are so plentiful.

Let us briefly recall the Osmia's habits. Amid the tangle of a hedge, a bramble-stalk is selected, still standing, but a mere withered stump. In this the insect digs a more or less deep tunnel, an easy piece of work owing to the abundance of soft pith. Provisions are heaped up right at the bottom of the tunnel and an egg is laid on the surface of the food: that is the first-born of the family. At a height of some twelve millimetres (About half an inch.—Translator's Note.), a partition is fixed. This gives a second storey, which in its turn receives provisions and an egg, the second in order of primogeniture. And so it goes on, storey by storey, until the cylinder is full. Then the thick plug of the same green material of which the partitions are formed closes the home and keeps out marauders.

In this common cradle, the chronological order of births is perfectly clear. The first-born of the family is at the bottom of the series; the last-born is at the top, near the closed door. The others follow from bottom to top in the same order in which they followed in point of time. The laying is numbered automatically; each cocoon tells us its respective age by the place which it occupies.

A number of eggs bordering on fifteen represents the entire family of an Osmia, and my observations enable me to state that the distribution of the sexes is not governed by any rule. All that I can say in general is that the complete series begins with females and nearly always ends with males. The incomplete series— those which the insect has laid in various places—can teach us nothing in this respect, for they are only fragments starting we know not whence; and it is impossible to tell whether they should be ascribed to the beginning, to the end, or to an intermediate period of the laying. To sum up: in the laying of the Three-pronged Osmia, no order governs the succession of the sexes; only, the series has a marked tendency to begin with females and to finish with males.

The mother occupies herself at the start with the stronger sex, the more necessary, the better-gifted, the female sex, to which she devotes the first flush of her laying and the fullness of her vigour; later, when she is perhaps already at the end of her strength, she bestows what remains of her maternal solicitude upon the weaker sex, the less-gifted, almost negligible male sex. There are, however, other species where this law becomes absolute, constant and regular.

In order to go more deeply into this curious question I installed some hives of a new kind on the sunniest walls of my enclosure. They consisted of stumps of the great reed of the south, open at one end, closed at the other by the natural knot and gathered into a sort of enormous pan-pipe, such as Polyphemus might have employed. The invitation was accepted: Osmiae came in fairly large numbers, to benefit by the queer installation.

Three Osmiae especially (O. Tricornis, Latr., O. cornuta, Latr., O. Latreillii, Spin.) gave me splendid results, with reed-stumps arranged either against the wall of my garden, as I have just said, or near their customary abode, the huge nests of the Mason-bee of the Sheds. One of them, the Three-horned Osmia, did better still: as I have described, she built her nests in my study, as plentifully as I could wish.

We will consult this last, who has furnished me with documents beyond my fondest hopes, and begin by asking her of how many eggs her average laying consists.

Of the whole heap of colonized tubes in my study, or else out of doors, in the hurdle-reeds and the pan-pipe appliances, the best-filled contains fifteen cells, with a free space above the series, a space showing that the laying is ended, for, if the mother had any more eggs available, she would have lodged them in the room which she leaves unoccupied. This string of fifteen appears to be rare; it was the only one that I found. My attempts at indoor rearing, pursued during two years with glass tubes or reeds, taught me that the Three-horned Osmia is not much addicted to long series. As though to decrease the difficulties of the coming deliverance, she prefers short galleries, in which only a part of the laying is stacked. We must then follow the same mother in her migration from one dwelling to the next if we would obtain a complete census of her family. A spot of colour, dropped on the Bee's thorax with a paint-brush while she is absorbed in closing up the mouth of the tunnel, enables us to recognize the Osmia in her various homes.

In this way, the swarm that resided in my study furnished me, in the first year, with an average of twelve cells. Next year, the summer appeared to be more favourable and the average became rather higher, reaching fifteen. The most numerous laying performed under my eyes, not in a tube, but in a succession of Snail-shells, reached the figure of twenty-six. On the other hand, layings of between eight and ten are not uncommon. Lastly, taking all my records together, the result is that the family of the Osmia fluctuates roundabout fifteen in number.

I have already spoken of the great differences in size apparent in the cells of one and the same series. The partitions, at first widely spaced, draw gradually nearer to one another as they come closer to the aperture, which implies roomy cells at the back and narrow cells in front. The contents of these compartments are no less uneven between one portion and another of the string. Without any exception known to me, the large cells, those with which the series starts, have more abundant provisions than the straitened cells with which the series ends. The heap of honey and pollen in the first is twice or even thrice as large as that in the second. In the last cells, the most recent in date, the victuals are but a pinch of pollen, so niggardly in amount that we wonder what will become of the larva with that meagre ration.

One would think that the Osmia, when nearing the end of the laying, attaches no importance to her last-born, to whom she doles out space and food so sparingly. The first-born receive the benefit of her early enthusiasm: theirs is the well-spread table, theirs the spacious apartments. The work has begun to pall by the time that the last eggs are laid; and the last-comers have to put up with a scurvy portion of food and a tiny corner.

The difference shows itself in another way after the cocoons are spun. The large cells, those at the back, receive the bulky cocoons; the small ones, those in front, have cocoons only half or a third as big. Before opening them and ascertaining the sex of the Osmia inside, let us wait for the transformation into the perfect insect, which will take place towards the end of summer. If impatience get the better of us, we can open them at the end of July or in August. The insect is then in the nymphal stage; and it is easy, under this form, to distinguish the two sexes by the length of the antennae, which are larger in the males, and by the glassy protuberances on the forehead, the sign of the future armour of the females. Well, the small cocoons, those in the narrow front cells, with their scanty store of provisions, all belong to males; the big cocoons, those in the spacious and well-stocked cells at the back, all belong to females.

The conclusion is definite: the laying of the Three-horned Osmia consists of two distinct groups, first a group of females and then a group of males.

With my pan-pipe apparatus displayed on the walls of my enclosure and with old hurdle-reeds left lying flat out of doors, I obtained the Horned Osmia in fair quantities. I persuaded Latreille's Osmia to build her nest in reeds, which she did with a zeal which I was far from expecting. All that I had to do was to lay some reed-stumps horizontally within her reach, in the immediate neighbourhood of her usual haunts, namely, the nests of the Mason-bee of the Sheds. Lastly, I succeeded without difficulty in making her build her nests in the privacy of my study, with glass tubes for a house. The result surpassed my hopes.

With both these Osmiae, the division of the gallery is the same as with the Three-horned Osmia. At the back are large cells with plentiful provisions and widely-spaced partitions; in front, small cells, with scanty provisions and partitions close together. Also, the larger cells supplied me with big cocoons and females; the smaller cells gave me little cocoons and males. The conclusion therefore is exactly the same in the case of all three Osmiae.

These conclusions, as my notes show, apply likewise, in every respect, to the various species of Mason-bees; and one clear and simple rule stands out from this collection of facts. Apart from the strange exception of the Three-pronged Osmia, who mixes the sexes without any order, the Bees whom I studied and probably a crowd of others produce first a continuous series of females and then a continuous series of males, the latter with less provisions and smaller cells. This distribution of the sexes agrees with what we have long known of the Hive-bee, who begins her laying with a long sequence of workers, or sterile females, and ends it with a long sequence of males. The analogy continues down to the capacity of the cells and the quantities of provisions. The real females, the Queen-bees, have wax cells incomparably more spacious than the cells of the males and receive a much larger amount of food. Everything therefore demonstrates that we are here in the presence of a general rule.

OPTIONAL DETERMINATION OF THE SEXES.

But does this rule express the whole truth? Is there nothing beyond a laying in two series? Are the Osmiae, the Chalicodomae and the rest of them fatally bound by this distribution of the sexes into two distinct groups, the male group following upon the female group, without any mixing of the two? Is the mother absolutely powerless to make a change in this arrangement, should circumstances require it?

The Three-pronged Osmia already shows us that the problem is far from being solved. In the same bramble-stump, the two sexes occur very irregularly, as though at random. Why this mixture in the series of cocoons of a Bee closely related to the Horned Osmia and the Three-horned Osmia, who stack theirs methodically by separate sexes in the hollow of a reed? What the Bee of the brambles does cannot her kinswomen of the reeds do too? Nothing, so far as I know, explains this fundamental difference in a physiological act of primary importance. The three Bees belong to the same genus; they resemble one another in general outline, internal structure and habits; and, with this close similarity, we suddenly find a strange dissimilarity.

There is just one thing that might possibly arouse a suspicion of the cause of this irregularity in the Three-pronged Osmia's laying. If I open a bramble-stump in the winter to examine the Osmia's nest, I find it impossible, in the vast majority of cases, to distinguish positively between a female and a male cocoon: the difference in size is so small. The cells, moreover, have the same capacity: the diameter of the cylinder is the same throughout and the partitions are almost always the same distance apart. If I open it in July, the victualling-period, it is impossible for me to distinguish between the provisions destined for the males and those destined for the females. The measurement of the column of honey gives practically the same depth in all the cells. We find an equal quantity of space and food for both sexes.

This result makes us foresee what a direct examination of the two sexes in the adult form tells us. The male does not differ materially from the female in respect of size. If he is a trifle smaller, it is scarcely noticeable, whereas, in the Horned Osmia and the Three-horned Osmia, the male is only half or a third the size of the female, as we have seen from the respective bulk of their cocoons. In the Mason-bee of the Walls there is also a difference in size, though less pronounced.

The Three-pronged Osmia has not therefore to trouble about adjusting the dimensions of the dwelling and the quantity of the food to the sex of the egg which she is about to lay; the measure is the same from one end of the series to the other. It does not matter if the sexes alternate without order: one and all will find what they need, whatever their position in the row. The two other Osmiae, with their great disparity in size between the two sexes, have to be careful about the twofold consideration of board and lodging.

The more I thought about this curious question, the more probable it appeared to me that the irregular series of the Three-pronged Osmia and the regular series of the other Osmiae and of the Bees in general were all traceable to a common law. It seemed to me that the arrangement in a succession first of females and then of males did not account for everything. There must be something more. And I was right: that arrangement in series is only a tiny fraction of the reality, which is remarkable in a very different way. This is what I am going to prove by experiment.

The succession first of females and then of males is not, in fact, invariable. Thus, the Chalicodoma, whose nests serve for two or three generations, ALWAYS lays male eggs in the old male cells, which can be recognized by their lesser capacity, and female eggs in the old female cells of more spacious dimensions.

This presence of both sexes at a time, even when there are but two cells free, one spacious and the other small, proves in the plainest fashion that the regular distribution observed in the complete nests of recent production is here replaced by an irregular distribution, harmonizing with the number and holding-capacity of the chambers to be stocked. The Mason-bee has before her, let me suppose, only five vacant cells: two larger and three smaller. The total space at her disposal would do for about a third of the laying. Well, in the two large cells, she puts females; in the three small cells she puts males.

As we find the same sort of thing in all the old nests, we must needs admit that the mother knows the sex of the eggs which she is going to lay, because that egg is placed in a cell of the proper capacity. We can go further, and admit that the mother alters the order of succession of the sexes at her pleasure, because her layings, between one old nest and another, are broken up into small groups of males and females according to the exigencies of space in the actual nest which she happens to be occupying.

Here then is the Chalicodoma, when mistress of an old nest of which she has not the power to alter the arrangement, breaking up her laying into sections comprising both sexes just as required by the conditions imposed upon her. She there-fore decides the sex of the egg at will, for, without this prerogative, she could not, in the chambers of the nest which she owes to chance, deposit unerringly the sex for which those chambers were originally built; and this happens however small the number of chambers to be filled.

When the mother herself founds the dwelling, when she lays the first rows of bricks, the females come first and the males at the finish. But, when she is in the presence of an old nest, of which she is quite unable to alter the general arrange-ment, how is she to make use of a few vacant rooms, the large and small alike, if the sex of the egg be already irrevocably fixed? She can only do so by abandoning the arrangement in two consecutive rows and accommodating her laying to the varied exigencies of the home. Either she finds it impossible to make an econom-ical use of the old nest, a theory refuted by the evidence, or else she determines at will the sex of the egg which she is about to lay.

The Osmiae themselves will furnish the most conclusive evidence on the latter point. We have seen that these Bees are not generally miners, who themselves dig out the foundation of their cells. They make use of the old structures of others, or else of natural retreats, such as hollow stems, the spirals of empty shells and var-ious hiding-places in walls, clay or wood. Their work is confined to repairs to the house, such as partitions and covers. There are plenty of these retreats; and the insects would always find first-class ones if it thought of going any distance to look for them. But the Osmia is a stay-at-home: she returns to her birthplace and clings to it with a patience extremely difficult to exhaust. It is here, in this little familiar corner, that she prefers to settle her progeny. But then the apartments are few in number and of all shapes and sizes. There are long and short ones, spacious ones and narrow. Short of expatriating herself, a Spartan course, she has to use them all, from first to last, for she has no choice. Guided by these considerations, I embarked on the experiments which I will now describe.

I have said how my study became a populous hive, in which the Three-horned Osmia built her nests in the various appliances which I had prepared for her. Among these appliances, tubes, either of glass or reed, predominated. There were tubes of all lengths and widths. In the long tubes, entire or almost entire layings, with a series of females followed by a series of males, were deposited. As I have already referred to this result, I will not discuss it again. The short tubes were suf-ficiently varied in length to lodge one or other portion of the total laying. Basing my calculations on the respective lengths of the cocoons of the two sexes, on the thickness of the partitions and the final lid, I shortened some of these to the exact dimensions required for two cocoons only, of different sexes.

Well, these short tubes, whether of glass or reed, were seized upon as eagerly as the long tubes. Moreover, they yielded this splendid result: their contents, only a part of the total laying, always began with female and ended with male cocoons. This order was invariable; what varied was the number of cells in the long tubes and the proportion between the two sorts of cocoons, sometimes males predominating and sometimes females.

When confronted with tubes too small to receive all her family, the Osmia is in the same plight as the Mason-bee in the presence of an old nest. She thereupon acts exactly as the Chalicodoma does. She breaks up her laying, divides it into series as short as the room at her disposal demands; and each series begins with females and ends with males. This breaking up, on the one hand, into sections in all of which both sexes are represented and the division, on the other hand, of the entire laying into just two groups, one female, the other male, when the length of the tube permits, surely provide us with ample evidence of the insect's power to regulate the sex of the egg according to the exigencies of space.

And besides the exigencies of space one might perhaps venture to add those connected with the earlier development of the males. These burst their cocoons a couple of weeks or more before the females; they are the first who hasten to the sweets of the almond-tree. In order to release themselves and emerge into the glad sunlight without disturbing the string of cocoons wherein their sisters are still sleeping, they must occupy the upper end of the row; and this, no doubt, is the reason that makes the Osmia end each of her broken layings with males. Being next to the door, these impatient ones will leave the home without upsetting the shells that are slower in hatching.

I had offered at the same time to the Osmiae in my study some old nests of the Mason-bee of the Shrubs, which are clay spheroids with cylindrical cavities in them. These cavities are formed, as in the old nests of the Mason-bee of the Pebbles, of the cell properly so-called and of the exit-way which the perfect insect cut through the outer coating at the time of its deliverance. The diameter is about 7 millimetres (.273 inch.—Translator's Note.); their depth at the centre of the heap is 23 millimetres (.897 inch.—Translator's Note.) and at the edge averages 14 millimetres. (.546 inch.—Translator's Note.)

The deep central cells receive only the females of the Osmia; sometimes even the two sexes together, with a partition in the middle, the female occupying the lower and the male the upper storey. Lastly, the deeper cavities on the circumference are allotted to females and the shallower to males.

We know that the Three-horned Osmia prefers to haunt the habitations of the Bees who nidify in populous colonies, such as the Mason-bee of the Sheds and the Hairy-footed Anthophora, in whose nests I have noted similar facts.

Thus the sex of the egg is optional. The choice rests with the mother, who is guided by considerations of space and, according to the accommodation at her disposal, which is frequently fortuitous and incapable of modification, places a female in this cell and a male in that, so that both may have a dwelling of a size suited to their unequal development. This is the unimpeachable evidence of the numerous and varied facts which I have set forth. People unfamiliar with insect anatomy—the public for whom I write—would probably give the following explanation of this marvellous prerogative of the Bee: the mother has at her disposal a certain number of eggs, some of which are irrevocably female and the others irrevocably male: she is able to pick out of either group the one which she wants at the actual moment; and her choice is decided by the holding capacity of the cell that has to be stocked. Everything would then be limited to a judicious selection from the heap of eggs.

Should this idea occur to him, the reader must hasten to reject it. Nothing could be more false, as the most casual reference to anatomy will show. The female reproductive apparatus of the Hymenoptera consists generally of six ovarian tubes, something like glove-fingers, divided into bunches of three and ending in a common canal, the oviduct, which carries the eggs outside. Each of these glove-fingers is fairly wide at the base, but tapers sharply towards the tip, which is closed. It contains, arranged in a row, one after the other, like beads on a string, a certain number of eggs, five or six for instance, of which the lower ones are more or less developed, the middle ones halfway towards maturity, and the upper ones very rudimentary. Every stage of evolution is here represented, distributed regularly from bottom to top, from the verge of maturity to the vague outlines of the embryo. The sheath clasps its string of ovules so closely that any inversion of the order is impossible. Besides, an inversion would result in a gross absurdity: the replacing of a riper egg by another in an earlier stage of development.

Therefore, in each ovarian tube, in each glove-finger, the emergence of the eggs occurs according to the order governing their arrangement in the common sheath; and any other sequence is absolutely impossible. Moreover, at the nesting-period, the six ovarian sheaths, one by one and each in its turn, have at their base an egg which in a very short time swells enormously. Some hours or even a day before the laying, that egg by itself represents or even exceeds in bulk the whole of the ovigerous apparatus. This is the egg which is on the point of being laid. It is about to descend into the oviduct, in its proper order, at its proper time; and the mother has no power to make another take its place. It is this egg, necessarily this egg and no other, that will presently be laid upon the provisions, whether these be a mess of honey or a live prey; it alone is ripe, it alone lies at the entrance to the oviduct; none of the others, since they are farther back in the row and not at the right stage of development, can be substituted at this crisis. Its birth is inevitable.

What will it yield, a male or a female? No lodging has been prepared, no food collected for it; and yet both food and lodging have to be in keeping with the sex that will proceed from it. And here is a much more puzzling condition: the sex of that egg, whose advent is predestined, has to correspond with the space which the mother happens to have found for a cell. There is therefore no room for hesitation, strange though the statement may appear: the egg, as it descends from its ovarian tube, has no determined sex. It is perhaps during the few hours of its rapid development at the base of its ovarian sheath, it is perhaps on its passage through the oviduct that it receives, at the mother's pleasure, the final impress that will produce, to match the cradle which it has to fill, either a female or a male.

PERMUTATIONS OF SEX.

Thereupon the following question presents itself. Let us admit that, when the normal conditions remain, a laying would have yielded m females and n males. Then, if my conclusions are correct, it must be in the mother's power, when the conditions are different, to take from the m group and increase the n group to the same extent; it must be possible for her laying to be represented as m - 1, m - 2, m - 3, etc. females and by n + 1, n + 2, n + 3, etc. males, the sum of m + n remaining constant, but one of the sexes being partly permuted into the other. The ultimate conclusion even cannot be disregarded: we must admit a set of eggs represented by m - m, or zero, females and of n + m males, one of the sexes being completely replaced by the other. Conversely, it must be possible for the feminine series to be augmented from the masculine series to the extent of absorbing it entirely. It was to solve this question and some others connected with it that I undertook, for the second time, to rear the Three-horned Osmia in my study.

The problem on this occasion is a more delicate one; but I am also better-equipped. My apparatus consists of two small closed packing-cases, with the front side of each pierced with forty holes, in which I can insert my glass tubes and keep them in a horizontal position. I thus obtain for the Bees the darkness and mystery which suit their work and for myself the power of withdrawing from my hive, at any time, any tube that I wish, with the Osmia inside, so as to carry it to the light and follow, if need be with the aid of the lens, the operations of the busy worker. My investigations, however frequent and minute, in no way hinder the peaceable Bee, who remains absorbed in her maternal duties.

I mark a plentiful number of my guests with a variety of dots on the thorax, which enables me to follow any one Osmia from the beginning to the end of her laying. The tubes and their respective holes are numbered; a list, always lying open on my desk, enables me to note from day to day, sometimes from hour to hour, what happens in each tube and particularly the actions of the Osmiae whose backs bear distinguishing marks. As soon as one tube is filled, I replace it by another. Moreover, I have scattered in front of either hive a few handfuls of empty Snail-shells, specially chosen for the object which I have in view. Reasons which I will explain later led me to prefer the shells of Helix caespitum. Each of the shells, as and when stocked, received the date of the laying and the alphabetical sign corresponding with the Osmia to whom it belonged. In this way, I spent five or six weeks in continual observation. To succeed in an enquiry, the first and foremost condition is patience. This condition I fulfilled; and it was rewarded with the success which I was justified in expecting.

The tubes employed are of two kinds. The first, which are cylindrical and of the same width throughout, will be of use for confirming the facts observed in the first year of my experiments in indoor rearing. The others, the majority, consist of two cylinders which are of very different diameters, set end to end. The front cylinder, the one which projects a little way outside the hive and forms the entrance-hole, varies in width between 8 and 12 millimetres. (Between .312 and .468 inch.—Translator's Note.) The second, the back one, contained entirely within my packing-case, is closed at its far end and is 5 to 6 millimetres in diameter. (.195 to .234 inch.—Translator's Note.) Each of the two parts of the double-galleried tunnel, one narrow and one wide, measures at most a decimetre in length. (3.9 inches.—Translator's Note.) I thought it advisable to have these short tubes, as the Osmia is thus compelled to select different lodgings, each of them being insufficient in itself to accommodate the total laying. In this way I shall obtain a greater variety in the distribution of the sexes. Lastly, at the mouth of each tube, which projects slightly outside the case, there is a little paper tongue, forming a sort of perch on which the Osmia alights on her arrival and giving easy access to the house. With these facilities, the swarm colonized fifty-two double-galleried tubes, thirty-seven cylindrical tubes, seventy-eight Snail-shells and a few old nests of the Mason-bee of the Shrubs. From this rich mine of material I will take what I want to prove my case.

Every series, even when incomplete, begins with females and ends with males. To this rule I have not yet found an exception, at least in galleries of normal diame-

ter. In each new abode the mother busies herself first of all with the more important sex. Bearing this point in mind, would it be possible for me, by manoeuvring,
to obtain an inversion of this order and make the laying begin with males? I think
so, from the results already ascertained and the irresistible conclusions to be
drawn from them. The double-galleried tubes are installed in order to put my
conjectures to the proof.

The back gallery, 5 or 6 millimetres wide (.195 to .234 inch.—Translator's Note.),
is too narrow to serve as a lodging for normally developed females. If, therefore,
the Osmia, who is very economical of her space, wishes to occupy them, she will
be obliged to establish males there. And her laying must necessarily begin here,
because this corner is the rear-most part of the tube. The foremost gallery is wide,
with an entrance-door on the front of the hive. Here, finding the conditions to
which she is accustomed, the mother will go on with her laying in the order which
she prefers.

Let us now see what has happened. Of the fifty-two double-galleried tubes, about
a third did not have their narrow passage colonized. The Osmia closed its aperture communicating with the large passage; and the latter alone received the eggs.
This waste of space was inevitable. The female Osmiae, though nearly always larger than the males, present marked differences among one another: some are bigger, some are smaller. I had to adjust the width of the narrow galleries to Bees of
average dimensions. It may happen therefore that a gallery is too small to admit
the large-sized mothers to whom chance allots it. When the Osmia is unable to
enter the tube, obviously she will not colonize it. She then closes the entrance to
this space which she cannot use and does her laying beyond it, in the wide tube.
Had I tried to avoid these useless apparatus by choosing tubes of larger calibre, I
should have encountered another drawback: the medium-sized mothers, finding
themselves almost comfortable, would have decided to lodge females there. I had
to be prepared for it: as each mother selected her house at will and as I was unable
to interfere in her choice, a narrow tube would be colonized or not, according as
the Osmia who owned it was or was not able to make her way inside.

There remain some forty pairs of tubes with both galleries colonized. In these
there are two things to take into consideration. The narrow rear tubes of 5 or 5
1/2 millimetres (.195 to .214 inch.—Translator's Note.)—and these are the most
numerous—contain males and males only, but in short series, between one and
five. The mother is here so much hampered in her work that they are rarely occupied from end to end; the Osmia seems in a hurry to leave them and to go and
colonize the front tube, whose ample space will leave her the liberty of movement
necessary for her operations. The other rear tubes, the minority, whose diameter

is about 6 millimetres (.234 inch.—Translator's Note.), contain sometimes only females and sometimes females at the back and males towards the opening. One can see that a tube a trifle wider and a mother slightly smaller would account for this difference in the results. Nevertheless, as the necessary space for a female is barely provided in this case, we see that the mother avoids as far as she can a two-sex arrangement beginning with males and that she adopts it only in the last extremity. Finally, whatever the contents of the small tube may be, those of the large one, following upon it, never vary and consist of females at the back and males in front.

Though incomplete, because of circumstances very difficult to control, the result of the experiment is none the less very remarkable. Twenty-five apparatus contain only males in their narrow gallery, in numbers varying from a minimum of one to a maximum of five. After these comes the colony of the large gallery, beginning with females and ending with males. And the layings in these apparatus do not always belong to late summer or even to the intermediate period: a few small tubes contain the earliest eggs of the entire swarm. A couple of Osmiae, more forward than the others, set to work on the 23rd of April. Both of them started their laying by placing males in the narrow tubes. The meagre supply of provisions was enough in itself to show the sex, which proved later to be in accordance with my anticipations. We see then that, by my artifices, the whole swarm starts with the converse of the normal order. This inversion is continued, at no matter what period, from the beginning to the end of the operations. The series which, according to rule, would begin with females now begins with males. Once the larger gallery is reached, the laying is pursued in the usual order.

We have advanced one step and that no small one: we have seen that the Osmia, when circumstances require it, is capable of reversing the sequence of the sexes. Would it be possible, provided that the tube were long enough, to obtain a complete inversion, in which the entire series of the males should occupy the narrow gallery at the back and the entire series of the females the roomy gallery in front? I think not; and I will tell you why.

Long and narrow cylinders are by no means to the Osmia's taste, not because of their narrowness but because of their length. Observe that for each load of honey brought the worker is obliged to move backwards twice. She enters, head first, to begin by disgorging the honey-syrup from her crop. Unable to turn in a passage which she blocks entirely, she goes out backwards, crawling rather than walking, a laborious performance on the polished surface of the glass and a performance which, with any other surface, would still be very awkward, as the wings are bound to rub against the wall with their free end and are liable to get rumpled or bent. She goes out backwards, reaches the outside, turns round and goes in again, but this time the opposite way, so as to brush off the load of pollen from her abdomen on to the heap. If the gallery is at all long, this crawling backwards becomes troublesome after a time; and the Osmia soon abandons a passage that is too small to allow of free movement. I have said that the narrow tubes of my apparatus are, for the most part, only very incompletely colonized. The Bee, after lodging a small number of males in them, hastens to leave them. In the wide front gallery she can stay where she is and still be able to turn round easily for her different manipulations; she will avoid those two long journeys backwards, which are so exhausting and so bad for her wings.

Another reason no doubt prompts her not to make too great a use of the narrow passage, in which she would establish males, followed by females in the part where the gallery widens. The males have to leave their cells a couple of weeks or more before the females. If they occupy the back of the house they will die prisoners or else they will overturn everything on their way out. This risk is avoided by the order which the Osmia adopts.

In my tubes, with their unusual arrangement, the mother might well find the dilemma perplexing: there is the narrowness of the space at her disposal and there is the emergence later on. In the narrow tubes, the width is insufficient for the females; on the other hand, if she lodges males there, they are liable to perish, since they will be prevented from issuing at the proper moment. This would perhaps explain the mother's hesitation and her obstinacy in settling females in some of my apparatus which looked as if they could suit none but males.

A suspicion occurs to me, a suspicion aroused by my attentive examination of the narrow tubes. All, whatever the number of their inmates, are carefully plugged at the opening, just as separate tubes would be. It might therefore be the case that the narrow gallery at the back was looked upon by the Osmia not as the prolongation of the large front gallery, but as an independent tube. The facility with which the worker turns as soon as she reaches the wide tube, her liberty of action, which is now as great as in a doorway communicating with the outer air, might well be misleading and cause the Osmia to treat the narrow passage at the back as though the wide passage in front did not exist. This would account for the placing of the female in the large tube above the males in the small tube, an arrangement contrary to her custom.

I will not undertake to decide whether the mother really appreciates the danger of my snares, or whether she makes a mistake in considering only the space at her disposal and beginning with males, who are liable to remain imprisoned. At any rate, I perceive a tendency to deviate as little as possible from the order which safe-guards the emergence of both sexes. This tendency is demonstrated by her repugnance to colonizing my narrow tubes with long series of males. However, so far as we are concerned, it does not matter much what passes at such times in the Osmia's little brain. Enough for us to know that she dislikes narrow and long tubes, not because they are narrow, but because they are at the same time long.

And, in fact, she does very well with a short tube of the same diameter. Such are the cells in the old nests of the Mason-bee of the Shrubs and the empty shells of the Garden Snail. With the short tube the two disadvantages of the long tube are avoided. She has very little of that crawling backwards to do when she has a Snail-shell for the home of her eggs and scarcely any when the home is the cell of the Mason-bee. Moreover, as the stack of cocoons numbers two or three at most, the deliverance will be exempt from the difficulties attached to a long series. To per-suade the Osmia to nidify in a single tube long enough to receive the whole of her laying and at the same time narrow enough to leave her only just the possibility of admittance appears to me a project without the slightest chance of success: the Bee would stubbornly refuse such a dwelling or would content herself with entrusting only a very small portion of her eggs to it. On the other hand, with narrow but short cavities, success, without being easy, seems to me at least quite possible. Guided by these considerations, I embarked upon the most arduous part of my problem: to obtain the complete or almost complete permutation of one sex with the other; to produce a laying consisting only of males by offering the mother a series of lodgings suited only to males.

Let us in the first place consult the old nests of the Mason-bee of the Shrubs. I have said that these mortar spheroids, pierced all over with little cylindrical cavi-ties, are a adopted pretty eagerly by the Three-horned Osmia, who colonizes them before my eyes with females in the deep cells and males in the shallow cells. That is how things go when the old nest remains in its natural state. With a grater, however, I scrape the outside of another nest so as to reduce the depth of the cav-ities to some ten millimetres. (About two-fifths of an inch.—Translator's Note.) This leaves in each cell just room for one cocoon, surmounted by the closing stopper. Of the fourteen cavities in the nests, I leave two intact, measuring fifteen millimetres in depth. (.585 inch.—Translator's Note.) Nothing could be more striking than the result of this experiment, made in the first year of my home rear-ing. The twelve cavities whose depth had been reduced all received males; the two cavities left untouched received females.

A year passes and I repeat the experiment with a nest of fifteen cells; but this time all the cells are reduced to the minimum depth with the grater. Well, the fifteen cells, from first to last, are occupied by males. It must be quite understood that, in each case, all the offspring belonged to one mother, marked with her distinguishing dot and kept in sight as long as her laying lasted. He would indeed be difficult to please who refused to bow before the results of these two experiments. If, however, he is not yet convinced, here is something to remove his last doubts.

The Three-horned Osmia often settles her family in old shells, especially those of the Common Snail (Helix aspersa), who is so common under the stone-heaps and in the crevices of the little unmortared walls that support our terraces. In this species the spiral is wide open, so that the Osmia, penetrating as far down as the helical passage permits, finds, immediately above the point which is too narrow to pass, the space necessary for the cell of a female. This cell is succeeded by others, wider still, always for females, arranged in a line in the same way as in a straight tube. In the last whorl of the spiral, the diameter would be too great for a single row. Then longitudinal partitions are added to the transverse partitions, the whole resulting in cells of unequal dimensions in which males predominate, mixed with a few females in the lower storeys. The sequence of the sexes is therefore what it would be in a straight tube and especially in a tube with a wide bore, where the partitioning is complicated by subdivisions on the same level. A single Snail-shell contains room for six or eight cells. A large, rough earthen stopper finishes the nest at the entrance to the shell.

As a dwelling of this sort could show us nothing new, I chose for my swarm the Garden Snail (Helix caespitum), whose shell, shaped like a small swollen Ammonite, widens by slow degrees, the diameter of the usable portion, right up to the mouth, being hardly greater than that required by a male Osmia-cocoon. Moreover, the widest part, in which a female might find room, has to receive a thick stopping-plug, below which there will often be a free space. Under all these conditions, the house will hardly suit any but males arranged one after the other.

The collection of shells placed at the foot of each hive includes specimens of different sizes. The smallest are 18 millimetres (.7 inch.—Translator's Note.) in diameter and the largest 24 millimetres. (.936 inch.—Translator's Note.) There is room for two cocoons, or three at most, according to their dimensions.

Now these shells were used by my visitors without any hesitation, perhaps even with more eagerness than the glass tubes, whose slippery sides might easily be a little annoying to the Bee. Some of them were occupied on the first few days of the laying; and the Osmia who had started with a home of this sort would pass next to a second Snail-shell, in the immediate neighbourhood of the first, to a third, a fourth and others still, always close together, until her ovaries were emptied. The whole family of one mother would thus be lodged in Snail-shells which were duly marked with the date of the laying and a description of the worker. The faithful adherents of the Snail-shell were in the minority. The greater number left the tubes to come to the shells and then went back from the shells to the tubes. All, after filling the spiral staircase with two or three cells, closed the house with a thick earthen stopper on a level with the opening. It was a long and troublesome task, in which the Osmia displayed all her patience as a mother and all her talents as a plasterer.

When the pupae are sufficiently matured, I proceed to examine these elegant abodes. The contents fill me with joy: they fulfil my anticipations to the letter. The great, the very great majority of the cocoons turn out to be males; here and there, in the bigger cells, a few rare females appear. The smallness of the space has almost done away with the stronger sex. This result is demonstrated by the sixty-eight Snail-shells colonized. But, of this total number, I must use only those series which received an entire laying and were occupied by the same Osmia from the beginning to the end of the egg-season. Here are a few examples, taken from among the most conclusive.

From the 6th of May, when she started operations, to the 25th of May, the date at which her laying ceased, one Osmia occupied seven Snail-shells in succession. Her family consists of fourteen cocoons, a number very near the average; and, of these fourteen cocoons, twelve belong to males and only two to females.

Another, between the 9th and 27th of May, stocked six Snail-shells with a family of thirteen, including ten males and three females.

A third, between the 2nd and 29th of May colonized eleven Snail-shells, a prodigious task. This industrious one was also exceedingly prolific. She supplied me with a family of twenty-six, the largest which I have ever obtained from one Osmia. Well, this abnormal progeny consisted of twenty-five males and one female.

There is no need to go on, after this magnificent example, especially as the other series would all, without exception, give us the same result. Two facts are immediately obvious: the Osmia is able to reverse the order of her laying and to start with a more or less long series of males before producing any females. There is something better still; and this is the proposition which I was particularly anxious to prove: the female sex can be permuted with the male sex and can be permuted to the point of disappearing altogether. We see this especially in the third case, where the presence of a solitary female in a family of twenty-six is due to the somewhat larger diameter of the corresponding Snail-shell.

There would still remain the inverse permutation: to obtain only females and no males, or very few. The first permutation makes the second seem very probable, although I cannot as yet conceive a means of realizing it. The only condition which I can regulate is the dimensions of the home. When the rooms are small, the males abound and the females tend to disappear. With generous quarters, the converse would not take place. I should obtain females and afterwards an equal number of males, confined in small cells which, in case of need, would be bounded by numerous partitions. The factor of space does not enter into the question here. What artifice can we then employ to provoke this second permutation? So far, I can think of nothing that is worth attempting.

It is time to conclude. Leading a retired life, in the solitude of a village, having quite enough to do with patiently and obscurely ploughing my humble furrow, I know little about modern scientific views. In my young days I had a passionate longing for books and found it difficult to procure them; to-day, when I could almost have them if I wanted, I am ceasing to wish for them. It is what usually happens as life goes on. I do not therefore know what may have been done in the direction whither this study of the sexes has led me. If I am stating propositions that are really new or at least more comprehensive than the propositions already known, my words will perhaps sound heretical. No matter: as a simple translator of facts, I do not hesitate to make my statement, being fully persuaded that time will turn my heresy into orthodoxy. I will therefore recapitulate my conclusions.

Bees lay their eggs in series of first females and then males, when the two sexes are of different sizes and demand an unequal quantity of nourishment. When the two sexes are alike in size, as in the case of Latreille's Osmia, the same sequence may occur, but less regularly.

This dual arrangement disappears when the place chosen for the nest is not large enough to contain the entire laying. We then see broken layings, beginning with females and ending with males.

The egg, as it issues from the ovary, has not yet a fixed sex. The final impress that produces the sex is given at the moment of laying, or a little before.

So as to be able to give each larva the amount of space and food that suits it according as it is male or female, the mother can choose the sex of the egg which she is about to lay. To meet the conditions of the building, which is often the work of another or else a natural retreat that admits of little or no alteration, she lays either a male egg or a female egg AS SHE PLEASES. The distribution of the sexes depends upon herself. Should circumstances require it, the order of the laying can be reversed and begin with males; lastly, the entire laying can contain only one sex.

The same privilege is possessed by the predatory Hymenoptera, the Wasps, at least by those in whom the two sexes are of a different size and consequently require an amount of nourishment that is larger in the one case than in the other. The mother must know the sex of the egg which she is going to lay; she must be able to choose the sex of that egg so that each larva may obtain its proper portion of food.

Generally speaking, when the sexes are of different sizes, every insect that collects food and prepares or selects a dwelling for its offspring must be able to choose the sex of the egg in order to satisfy without mistake the conditions imposed upon it.

The question remains how this optional assessment of the sexes is effected. I know absolutely nothing about it. If I should ever learn anything about this delicate point, I shall owe it to some happy chance for which I must wait, or rather watch, patiently.

Then what explanation shall I give of the wonderful facts which I have set forth? Why, none, absolutely none. I do not explain facts, I relate them. Growing daily more sceptical of the interpretations suggested to me and more hesitating as to those which I myself may have to suggest, the more I observe and experiment, the more clearly I see rising out of the black mists of possibility an enormous note of interrogation.

Dear insects, my study of you has sustained me and continues to sustain me in my heaviest trials; I must take leave of you for to-day. The ranks are thinning around me and the long hopes have fled. Shall I be able to speak of you again? (This forms the closing paragraph of Volume 3 of the "Souvenirs entomologiques," of which the author lived to publish seven more volumes, containing over 2,500 pages and nearly 850,000 words.—Translator's Note.)

CHAPTER 13.

THE GLOW-WORM.

Few insects in our climes vie in popular fame with the Glow-worm, that curious little animal which, to celebrate the little joys of life, kindles a beacon at its tail-end. Who does not know it, at least by name? Who has not seen it roam amid the grass, like a spark fallen from the moon at its full? The Greeks of old called it lampouris, meaning, the bright-tailed. Science employs the same term: it calls it the lantern-bearer, Lampyris noctiluca, Lin. In this case the common name is inferior to the scientific phrase, which, when translated, becomes both expressive and accurate.

In fact, we might easily cavil at the word "worm." The Lampyris is not a worm at all, not even in general appearance. He has six short legs, which he well knows how to use; he is a gad-about, a trot-about. In the adult state the male is correctly garbed in wing-cases, like the true Beetle that he is. The female is an ill-favoured thing who knows naught of the delights of flying: all her life long she retains the larval shape, which, for the rest, is similar to that of the male, who himself is imperfect so long as he has not achieved the maturity that comes with pairing-time. Even in this initial stage the word "worm" is out of place. We French have the expression "Naked as a worm" to point to the lack of any defensive covering. Now the Lampyris is clothed, that is to say, he wears an epidermis of some consistency; moreover, he is rather richly coloured: his body is dark brown all over, set off with pale pink on the thorax, especially on the lower surface. Finally, each segment is decked at the hinder edge with two spots of a fairly bright red. A costume like this was never worn by a worm.

Let us leave this ill-chosen denomination and ask ourselves what the Lampyris feeds upon. That master of the art of gastronomy, Brillat-Savarin, said: "Show me what you eat and I will tell you what you are."

A similar question should be addressed, by way of a preliminary, to every insect whose habits we propose to study, for, from the least to the greatest in the zoological progression, the stomach sways the world; the data supplied by food are the chief of all the documents of life. Well, in spite of his innocent appearance, the Lampyris is an eater of flesh, a hunter of game; and he follows his calling with rare villainy. His regular prey is the Snail.

This detail has long been known to entomologists. What is not so well known, what is not known at all yet, to judge by what I have read, is the curious method of attack, of which I have seen no other instance anywhere.

Before he begins to feast, the Glow-worm administers an anaesthetic: he chloroforms his victim, rivalling in the process the wonders of our modern surgery, which renders the patient insensible before operating on him. The usual game is a small Snail hardly the size of a cherry, such as, for instance, Helix variabilis, Drap., who, in the hot weather, collects in clusters on the stiff stubble and other long, dry stalks by the road-side and there remains motionless, in profound meditation, throughout the scorching summer days. It is in some such resting-place as this that I have often been privileged to light upon the Lampyris banqueting on the prey which he had just paralysed on its shaky support by his surgical artifices.

But he is familiar with other preserves. He frequents the edges of the irrigating ditches, with their cool soil, their varied vegetation, a favourite haunt of the Mollusc. Here, he treats the game on the ground; and, under these conditions, it is easy for me to rear him at home and to follow the operator's performance down to the smallest detail.

I will try to make the reader a witness of the strange sight. I place a little grass in a wide glass jar. In this I instal a few Glow-worms and a provision of snails of a suitable size, neither too large nor too small, chiefly Helix variabilis. We must be patient and wait. Above all, we must keep an assiduous watch, for the desired events come unexpectedly and do not last long.

Here we are at last. The Glow-worm for a moment investigates the prey, which, according to its habit, is wholly withdrawn in the shell, except the edge of the mantle, which projects slightly. Then the hunter's weapon is drawn, a very simple

weapon, but one that cannot be plainly perceived without the aid of a lens. It consists of two mandibles bent back powerfully into a hook, very sharp and as thin
as a hair. The microscope reveals the presence of a slender groove running
throughout the length. And that is all.

The insect repeatedly taps the Snail's mantle with its instrument. It all happens
with such gentleness as to suggest kisses rather than bites. As children, teasing one
another, we used to talk of "tweaksies" to express a slight squeeze of the finger-
tips, something more like a tickling than a serious pinch. Let us use that word. In
conversing with animals, language loses nothing by remaining juvenile. It is the
right way for the simple to understand one another.

The Lampyris doles out his tweaks. He distributes them methodically, without
hurrying, and takes a brief rest after each of them, as though he wished to ascertain the effect produced. Their number is not great: half a dozen, at most, to subdue the prey and deprive it of all power of movement. That other pinches are
administered later, at the time of eating, seems very likely, but I cannot say anything for certain, because the sequel escapes me. The first few, however—there are
never many—are enough to impart inertia and loss of all feeling to the Mollusc,
thanks to the prompt, I might almost say lightning, methods of the Lampyris,
who, beyond a doubt, instils some poison or other by means of his grooved hooks.

Here is the proof of the sudden efficacy of those twitches, so mild in appearance:
I take the Snail from the Lampyris, who has operated on the edge of the mantle
some four or five times. I prick him with a fine needle in the fore-part, which the
animal, shrunk into its shell, still leaves exposed. There is no quiver of the wound-
ed tissues, no reaction against the brutality of the needle. A corpse itself could not
give fewer signs of life.

Here is something even more conclusive: chance occasionally gives me Snails
attacked by the Lampyris while they are creeping along, the foot slowly crawling,
the tentacles swollen to their full extent. A few disordered movements betray a
brief excitement on the part of the Mollusc and then everything ceases: the foot
no longer slugs; the front part loses its graceful swan-neck curve; the tentacles
become limp and give way under their own weight, dangling feebly like a broken
stick. This condition persists.

Is the Snail really dead? Not at all, for I can resuscitate the seeming corpse at will.
After two or three days of that singular condition which is no longer life and yet
not death, I isolate the patient and, though this is not really essential to success,
I give him a douche which will represent the shower so dear to the able-bodied

Mollusc. In about a couple of days, my prisoner, but lately injured by the Glow-worm's treachery, is restored to his normal state. He revives, in a manner; he recovers movement and sensibility. He is affected by the stimulus of a needle; he shifts his place, crawls, puts out his tentacles, as though nothing unusual had occurred. The general torpor, a sort of deep drunkenness, has vanished outright. The dead returns to life. What name shall we give to that form of existence which, for a time, abolishes the power of movement and the sense of pain? I can see but one that is approximately suitable: anaesthesia. The exploits of a host of Wasps whose flesh-eating grubs are provided with meat that is motionless though not dead have taught us the skilful art of the paralysing insect, which numbs the loco-motory nerve-centres with its venom. We have now a humble little animal that first produces complete anaesthesia in its patient. Human science did not in real-ity invent this art, which is one of the wonders of latter-day surgery. Much earli-er, far back in the centuries, the Lampyris and, apparently, others knew it as well. The animal's knowledge had a long start of ours; the method alone has changed. Our operators proceed by making us inhale the fumes of ether or chloroform; the insect proceeds by injecting a special virus that comes from the mandibular fangs in infinitesimal doses. Might we not one day be able to benefit from this hint? What glorious discoveries the future would have in store for us, if we understood the beastie's secrets better!

What does the Lampyris want with anaesthetical talent against a harmless and moreover eminently peaceful adversary, who would never begin the quarrel of his own accord? I think I see. We find in Algeria a beetle known as Drilus maroc-canus, who, though non-luminous, approaches our Glow-worm in his organiza-tion and especially in his habits. He, too, feeds on Land Molluscs. His prey is a Cyclostome with a graceful spiral shell, tightly closed with a stony lid which is attached to the animal by a powerful muscle. The lid is a movable door which is quickly shut by the inmate's mere withdrawal into his house and as easily opened when the hermit goes forth. With this system of closing, the abode becomes invi-olable; and the Drilus knows it.

Fixed to the surface of the shell by an adhesive apparatus whereof the Lampyris will presently show us the equivalent, he remains on the look-out, waiting, if nec-essary, for whole days at a time. At last the need of air and food obliges the besieged non-combatant to show himself: at least, the door is set slightly ajar. That is enough. The Drilus is on the spot and strikes his blow. The door can no longer be closed; and the assailant is henceforth master of the fortress. Our first impression is that the muscle moving the lid has been cut with a quick-acting pair of shears. This idea must be dismissed. The Drilus is not well enough equipped with jaws to gnaw through a fleshy mass so promptly. The operation has to succeed at once, at the first touch: if not, the animal attacked would retreat, still in full vigour, and the siege must be recommenced, as arduous as ever, exposing the insect to fasts indefinitely prolonged. Although I have never come across the Drilus, who is a stranger to my district, I conjecture a method of attack very similar to that of the Glow-worm. Like our own Snail-eater, the Algerian insect does not cut its victim into small pieces: it renders it inert, chloroforms it by means of a few tweaks which are easily distributed, if the lid but half-opens for a second. That will do. The besieger thereupon enters and, in perfect quiet, consumes a prey incapable of the least muscular effort. That is how I see things by the unaided light of logic.

Let us now return to the Glow-worm. When the Snail is on the ground, creeping, or even shrunk into his shell, the attack never presents any difficulty. The shell possesses no lid and leaves the hermit's fore-part to a great extent exposed. Here, on the edges of the mantle, contracted by the fear of danger, the Mollusc is vulnerable and incapable of defence. But it also frequently happens that the Snail occupies a raised position, clinging to the tip of a grass-stalk or perhaps to the smooth surface of a stone. This support serves him as a temporary lid; it wards off the aggression of any churl who might try to molest the inhabitant of the cabin, always on the express condition that no slit show itself anywhere on the protecting circumference. If, on the other hand, in the frequent case when the shell does not fit its support quite closely, some point, however tiny, be left uncovered, this is enough for the subtle tools of the Lampyris, who just nibbles at the Mollusc and at once plunges him into that profound immobility which favours the tranquil proceedings of the consumer.

These proceedings are marked by extreme prudence. The assailant has to handle his victim gingerly, without provoking contractions which would make the Snail let go his support and, at the very least, precipitate him from the tall stalk whereon he is blissfully slumbering. Now any game falling to the ground would seem to be so much sheer loss, for the Glow-worm has no great zeal for hunting-expeditions: he profits by the discoveries which good luck sends him, without undertaking assiduous searches. It is essential, therefore, that the equilibrium of a prize perched on the top of a stalk and only just held in position by a touch of glue should be disturbed as little as possible during the onslaught; it is necessary that the assailant should go to work with infinite circumspection and without producing pain, lest any muscular reaction should provoke a fall and endanger the prize. As we see, sudden and profound anaesthesia is an excellent means of enabling the Lampyris to attain his object, which is to consume his prey in perfect quiet.

What is his manner of consuming it? Does he really eat, that is to say, does he divide his food piecemeal, does he carve it into minute particles, which are afterwards ground by a chewing-apparatus? I think not. I never see a trace of solid nourishment on my captives' mouths. The Glow-worm does not eat in the strict sense of the word: he drinks his fill; he feeds on a thin gruel into which he transforms his prey by a method recalling that of the maggot. Like the flesh-eating grub of the Fly, he too is able to digest before consuming; he liquefies his prey before feeding on it.

This is how things happen: a Snail has been rendered insensible by the Glow-worm. The operator is nearly always alone, even when the prize is a large one, like the common Snail, Helix aspersa. Soon a number of guests hasten up—two, three, or more—and, without any quarrel with the real proprietor, all alike fall to. Let us leave them to themselves for a couple of days and then turn the shell, with the opening downwards. The contents flow out as easily as would soup from an overturned saucepan. When the sated diners retire from this gruel, only insignificant leavings remain.

The matter is obvious. By repeated tiny bites, similar to the tweaks which we saw distributed at the outset, the flesh of the Mollusc is converted into a gruel on which the various banqueters nourish themselves without distinction, each working at the broth by means of some special pepsine and each taking his own mouthfuls of it. In consequence of this method, which first converts the food into a liquid, the Glow-worm's mouth must be very feebly armed apart from the two fangs which sting the patient and inject the anaesthetic poison and at the same time, no doubt, the serum capable of turning the solid flesh into fluid. Those two tiny implements, which can just be examined through the lens, must, it seems, have some other object. They are hollow, and in this resemble those of the Ant-lion, who sucks and drains her capture without having to divide it; but there is this great difference, that the Ant-lion leaves copious remnants, which are afterwards flung outside the funnel-shaped trap dug in the sand, whereas the Glow-worm, that expert liquefier, leaves nothing, or next to nothing. With similar tools, the one simply sucks the blood of his prey and the other turns every morsel of his to account, thanks to a preliminary liquefaction.

And this is done with exquisite precision, though the equilibrium is sometimes anything but steady. My rearing-glasses supply me with magnificent examples. Crawling up the sides, the Snails imprisoned in my apparatus sometimes reach the top, which is closed with a glass pane, and fix themselves to it with a speck of glair. This is a mere temporary halt, in which the Mollusc is miserly with his adhesive product, and the merest shake is enough to loosen the shell and send it to the bottom of the jar.

Now it is not unusual for the Glow-worm to hoist himself up there, with the help of a certain climbing-organ that makes up for his weak legs. He selects his quarry, makes a minute inspection of it to find an entrance-slit, nibbles at it a little, renders it insensible and, without delay, proceeds to prepare the gruel which he will consume for days on end.

When he leaves the table, the shell is found to be absolutely empty; and yet this shell, which was fixed to the glass by a very faint stickiness, has not come loose, has not even shifted its position in the smallest degree: without any protest from the hermit gradually converted into broth, it has been drained on the very spot at which the first attack was delivered. These small details tell us how promptly the anaesthetic bite takes effect; they teach us how dexterously the Glow-worm treats his Snail without causing him to fall from a very slippery, vertical support and without even shaking him on his slight line of adhesion.

Under these conditions of equilibrium, the operator's short, clumsy legs are obviously not enough; a special accessory apparatus is needed to defy the danger of slipping and to seize the unseizable. And this apparatus the Lampyris possesses. At the hinder end of the animal we see a white spot which the lens separates into some dozen short, fleshy appendages, sometimes gathered into a cluster, sometimes spread into a rosette. There is your organ of adhesion and locomotion. If he would fix himself somewhere, even on a very smooth surface, such as a grass-stalk, the Glow-worm opens his rosette and spreads it wide on the support, to which it adheres by its own stickiness. The same organ, rising and falling, opening and closing, does much to assist the act of progression. In short, the Glow-worm is a new sort of self-propelled cripple, who decks his hind-quarters with a dainty white rose, a kind of hand with twelve fingers, not jointed, but moving in every direction: tubular fingers which do not seize, but stick.

The same organ serves another purpose: that of a toilet-sponge and brush. At a moment of rest, after a meal, the Glow-worm passes and repasses the said brush over his head, back, sides and hinder parts, a performance made possible by the flexibility of his spine. This is done point by point, from one end of the body to the other, with a scrupulous persistency that proves the great interest which he takes in the operation. What is his object in thus sponging himself, in dusting and polishing himself so carefully? It is a question, apparently, of removing a few atoms of dust or else some traces of viscidity that remain from the evil contact with the Snail. A wash and brush-up is not superfluous when one leaves the tub in which the Mollusc has been treated.

If the Glow-worm possessed no other talent than that of chloroforming his prey by means of a few tweaks resembling kisses, he would be unknown to the vulgar herd; but he also knows how to light himself like a beacon; he shines, which is an excellent manner of achieving fame. Let us consider more particularly the female, who, while retaining her larval shape, becomes marriageable and glows at her best during the hottest part of summer. The lighting-apparatus occupies the last three segments of the abdomen. On each of the first two it takes the form, on the ventral surface, of a wide belt covering almost the whole of the arch; on the third the luminous part is much less and consists simply of two small crescent-shaped markings, or rather two spots which shine through to the back and are visible both above and below the animal. Belts and spots emit a glorious white light, delicately tinged with blue. The general lighting of the Glow-worm thus comprises two groups: first, the wide belts of the two segments preceding the last; secondly, the two spots of the final segments. The two belts, the exclusive attribute of the marriageable female, are the parts richest in light: to glorify her wedding, the future mother dons her brightest gauds; she lights her two resplendent scarves. But, before that, from the time of the hatching, she had only the modest rush-light of the stern. This efflorescence of light is the equivalent of the final metamorphosis, which is usually represented by the gift of wings and flight. Its brilliance heralds the pairing-time. Wings and flight there will be none: the female retains her humble larval form, but she kindles her blazing beacon.

The male, on his side, is fully transformed, changes his shape, acquires wings and wing-cases; nevertheless, like the female, he possesses, from the time when he is hatched, the pale lamp of the end segment. This luminous aspect of the stern is characteristic of the entire Glow-worm tribe, independently of sex and season. It appears upon the budding grub and continues throughout life unchanged. And we must not forget to add that it is visible on the dorsal as well as on the ventral surface, whereas the two large belts peculiar to the female shine only under the abdomen.

My hand is not so steady nor my sight so good as once they were; but, as far as they allow me, I consult anatomy for the structure of the luminous organs. I take a scrap of the epidermis and manage to separate pretty nearly half of one of the shining belts. I place my preparation under the microscope. On the skin a sort of white-wash lies spread, formed of a very fine, granular substance. This is certainly the light-producing matter. To examine this white layer more closely is beyond the power of my weary eyes. Just beside it is a curious air-tube, whose short and remarkably wide stem branches suddenly into a sort of bushy tuft of very delicate ramifications. These creep over the luminous sheet, or even dip into it. That is all.

The luminescence, therefore, is controlled by the respiratory organs and the work produced is an oxidation. The white sheet supplies the oxidizable matter and the thick air-tube spreading into a tufty bush distributes the flow of air over it. There remains the question of the substance whereof this sheet is formed. The first suggestion was phosphorus, in the chemist's sense of the word. The Glow-worm was calcined and treated with the violent reagents that bring the simple substances to light; but no one, so far as I know, has obtained a satisfactory answer along these lines. Phosphorus seems to play no part here, in spite of the name of phosphorescence which is sometimes bestowed upon the Glow-worm's gleam. The answer lies elsewhere, no one knows where.

We are better-informed as regards another question. Has the Glow-worm a free control of the light which he emits? Can he turn it on or down or put it out as he pleases? Has he an opaque screen which is drawn over the flame at will, or is that flame always left exposed? There is no need for any such mechanism: the insect has something better for its revolving light.

The thick air-tube supplying the light-producing sheet increases the flow of air and the light is intensified; the same tube, swayed by the animal's will, slackens or even suspends the passage of air and the light grows fainter or even goes out. It is, in short, the mechanism of a lamp which is regulated by the access of air to the wick.

Excitement can set the attendant air-duct in motion. We must here distinguish between two cases: that of the gorgeous scarves, the exclusive ornament of the female ripe for matrimony, and that of the modest fairy-lamp on the last segment, which both sexes kindle at any age. In the second case, the extinction caused by a flurry is sudden and complete, or nearly so. In my nocturnal hunts for young Glow-worms, measuring about 5 millimetres long (.195 inch.—Translator's Note.), I can plainly see the glimmer on the blades of grass; but, should the least false step disturb a neighbouring twig, the light goes out at once and the coveted insect becomes invisible. Upon the full-grown females, lit up with their nuptial scarves, even a violent start has but a slight effect and often none at all.

I fire a gun beside a wire-gauze cage in which I am rearing my menagerie of females in the open air. The explosion produces no result. The illumination continues, as bright and placid as before. I take a spray and rain down a slight shower of cold water upon the flock. Not one of my animals puts out its light; at the very most, there is a brief pause in the radiance; and then only in some cases. I send a puff of smoke from my pipe into the cage. This time the pause is more marked. There are even some extinctions, but these do not last long. Calm soon returns and the light is renewed as brightly as ever. I take some of the captives in my fingers, turn and return them, tease them a little. The illumination continues and is not much diminished, if I do not press hard with my thumb. At this period, with the pairing close at hand, the insect is in all the fervour of its passionate splendour, and nothing short of very serious reasons would make it put out its signals altogether.

All things considered, there is not a doubt but that the Glow-worm himself manages his lighting apparatus, extinguishing and rekindling it at will; but there is one point at which the voluntary agency of the insect is without effect. I detach a strip of the epidermis showing one of the luminescent sheets and place it in a glass tube, which I close with a plug of damp wadding, to avoid an over-rapid evaporation. Well, this scrap of carcass shines away merrily, although not quite as brilliantly as on the living body.

Life's aid is now superfluous. The oxidizable substance, the luminescent sheet, is in direct communication with the surrounding atmosphere; the flow of oxygen through an air-tube is not necessary; and the luminous emission continues to take place, in the same way as when it is produced by the contact of the air with the real phosphorus of the chemists. Let us add that, in aerated water, the luminousness continues as brilliant as in the free air, but that it is extinguished in water deprived of its air by boiling. No better proof could be found of what I have already propounded, namely, that the Glow-worm's light is the effect of a slow oxidation.

The light is white, calm and soft to the eyes and suggests a spark dropped by the full moon. Despite its splendour, it is a very feeble illuminant. If we move a Glow-worm along a line of print, in perfect darkness, we can easily make out the letters, one by one, and even words, when these are not too long; but nothing more is visible beyond a narrow zone. A lantern of this kind soon tires the reader's patience.

Suppose a group of Glow-worms placed almost touching one another. Each of them sheds its glimmer, which ought, one would think, to light up its neighbours by reflexion and give us a clear view of each individual specimen. But not at all: the luminous party is a chaos in which our eyes are unable to distinguish any definite form at a medium distance. The collective lights confuse the light-bearers into one vague whole.

Photography gives us a striking proof of this. I have a score of females, all at the height of their splendour, in a wire-gauze cage in the open air. A tuft of thyme forms a grove in the centre of their establishment. When night comes, my captives clamber to this pinnacle and strive to show off their luminous charms to the best advantage at every point of the horizon, thus forming along the twigs marvellous clusters from which I expected magnificent effects on the photographer's plates and paper. My hopes were disappointed. All that I obtain is white, shapeless patches, denser here and less dense there according to the numbers forming the group. There is no picture of the Glow-worms themselves; not a trace either of the tuft of thyme. For want of satisfactory light, the glorious firework is represented by a blurred splash of white on a black ground.

The beacons of the female Glow-worms are evidently nuptial signals, invitations to the pairing; but observe that they are lighted on the lower surface of the abdomen and face the ground, whereas the summoned males, whose flights are sudden and uncertain, travel overhead, in the air, sometimes a great way up. In its normal position, therefore, the glittering lure is concealed from the eyes of those concerned; it is covered by the thick bulk of the bride. The lantern ought really to gleam on the back and not under the belly; otherwise the light is hidden under a bushel.

The anomaly is corrected in a very ingenious fashion, for every female has her little wiles of coquetry. At nightfall, every evening, my caged captives make for the tuft of thyme with which I have thoughtfully furnished the prison and climb to the top of the upper branches, those most in sight. Here, instead of keeping quiet, as they did at the foot of the bush just now, they indulge in violent exercises, twist the tip of their very flexible abdomen, turn it to one side, turn it to the other, jerk it in every direction. In this way, the searchlight cannot fail to gleam, at one moment or another, before the eyes of every male who goes a-wooing in the neighbourhood, whether on the ground or in the air.

It is very like the working of the revolving mirror used in catching Larks. If stationary, the little contrivance would leave the bird indifferent; turning and breaking up its light in rapid flashes, it excites it.

While the female Glow-worm has her tricks for summoning her swains, the male, on his side, is provided with an optical apparatus suited to catch from afar the least reflection of the calling signal. His corselet expands into a shield and overlaps his head considerably in the form of a peaked cap or a shade, the object of which appears to be to limit the field of vision and concentrate the view upon the luminous speck to be discerned. Under this arch are the two eyes, which are relatively enormous, exceedingly convex, shaped like a skull-cap and contiguous to the extent of leaving only a narrow groove for the insertion of the antennae. This double eye, occupying almost the whole face of the insect and contained in the cavern formed by the spreading peak of the corselet, is a regular Cyclops' eye.

At the moment of the pairing the illumination becomes much fainter, is almost extinguished; all that remains alight is the humble fairy-lamp of the last segment. This discreet night-light is enough for the wedding, while, all around, the host of nocturnal insects, lingering over their respective affairs, murmur the universal marriage-hymn. The laying follows very soon. The round, white eggs are laid, or rather strewn at random, without the least care on the mother's part, either on the more or less cool earth or on a blade of grass. These brilliant ones know nothing at all of family affection.

Here is a very singular thing: the Glow-worm's eggs are luminous even when still contained in the mother's womb. If I happen by accident to crush a female big with germs that have reached maturity, a shiny streak runs along my fingers, as though I had broken some vessel filled with a phosphorescent fluid. The lens shows me that I am wrong. The luminosity comes from the cluster of eggs forced out of the ovary. Besides, as laying-time approaches, the phosphorescence of the eggs is already made manifest through this clumsy midwifery. A soft opalescent light shines through the integument of the belly.

The hatching follows soon after the laying. The young of either sex have two little rush-lights on the last segment. At the approach of the severe weather they go down into the ground, but not very far. In my rearing-jars, which are supplied with fine and very loose earth, they descend to a depth of three or four inches at most. I dig up a few in mid-winter. I always find them carrying their faint stern-light. About the month of April they come up again to the surface, there to continue and complete their evolution.

From start to finish the Glow-worm's life is one great orgy of light. The eggs are luminous; the grubs likewise. The full-grown females are magnificent lighthouses, the adult males retain the glimmer which the grubs already possessed. We can understand the object of the feminine beacon; but of what use is all the rest of the pyrotechnic display? To my great regret, I cannot tell. It is and will be, for many a day to come, perhaps for all time, the secret of animal physics, which is deeper than the physics of the books.

CHAPTER 14.

THE CABBAGE-CATERPILLAR.

The cabbage of our modern kitchen-gardens is a semi-artificial plant, the produce of our agricultural ingenuity quite as much as of the niggardly gifts of nature. Spontaneous vegetation supplied us with the long-stalked, scanty-leaved, ill-smelling wilding, as found, according to the botanists, on the ocean cliffs. He had need of a rare inspiration who first showed faith in this rustic clown and proposed to improve it in his garden-patch.

Progressing by infinitesimal degrees, culture wrought miracles. It began by per-suading the wild cabbage to discard its wretched leaves, beaten by the sea-winds, and to replace them by others, ample and fleshy and close-fitting. The gentle cab-bage submitted without protest. It deprived itself of the joys of light by arranging its leaves in a large compact head, white and tender. In our day, among the suc-cessors of those first tiny hearts, are some that, by virtue of their massive bulk, have earned the glorious name of chou quintal, as who should say a hundred-weight of cabbage. They are real monuments of green stuff.

Later, man thought of obtaining a generous dish with a thousand little sprays of the inflorescence. The cabbage consented. Under the cover of the central leaves, it gorged with food its sheaves of blossom, its flower-stalks, its branches and worked the lot into a fleshy conglomeration. This is the cauliflower, the broccoli.

Differently entreated, the plant, economizing in the centre of its shoot, set a whole family of close-wrapped cabbages ladder-wise on a tall stem. A multitude of dwarf leaf-buds took the place of the colossal head. This is the Brussels sprout.

Next comes the turn of the stump, an unprofitable, almost wooden, thing, which seemed never to have any other purpose than to act as a support for the plant. But the tricks of gardeners are capable of everything, so much so that the stalk yields to the grower's suggestions and becomes fleshy and swells into an ellipse similar to the turnip, of which it possesses all the merits of corpulence, flavour and delicacy; only the strange product serves as a base for a few sparse leaves, the last protests of a real stem that refuses to lose its attributes entirely. This is the cole-rape.

If the stem allows itself to be allured, why not the root? It does, in fact, yield to the blandishments of agriculture: it dilates its pivot into a flat turnip, which half emerges from the ground. This is the rutabaga, or swede, the turnip-cabbage of our northern districts.

Incomparably docile under our nursing, the cabbage has given its all for our nourishment and that of our cattle: its leaves, its flowers, its buds, its stalk, its root; all that it now wants is to combine the ornamental with the useful, to smarten itself, to adorn our flowerbeds and cut a good figure on a drawing-room table. It has done this to perfection, not with its flowers, which, in their modesty, continue intractable, but with its curly and variegated leaves, which have the undulating grace of Ostrich-feathers and the rich colouring of a mixed bouquet. None who beholds it in this magnificence will recognize the near relation of the vulgar "greens" that form the basis of our cabbage-soup.

The cabbage, first in order of date in our kitchen-gardens, was held in high esteem by classic antiquity, next after the bean and, later, the pea; but it goes much farther back, so far indeed that no memories of its acquisition remain. History pays but little attention to these details: it celebrates the battle-fields whereon we meet our death, but scorns to speak of the ploughed fields whereby we thrive; it knows the names of the kings' bastards, but cannot tell us the origin of wheat. That is the way of human folly.

This silence respecting the precious plants that serve as food is most regrettable. The cabbage in particular, the venerable cabbage, that denizen of the most ancient garden-plots, would have had extremely interesting things to teach us. It is a treasure in itself, but a treasure twice exploited, first by man and next by the caterpillar of the Pieris, the common Large White Butterfly whom we all know (Pieris brassicae, Lin.). This caterpillar feeds indiscriminately on the leaves of all varieties of cabbage, however dissimilar in appearance: he nibbles with the same appetite red cabbage and broccoli, curly greens and savoy, swedes and turnip-tops, in short, all that our ingenuity, lavish of time and patience, has been able to obtain from the original plant since the most distant ages.

But what did the caterpillar eat before our cabbages supplied him with copious provender? Obviously the Pieris did not wait for the advent of man and his horticultural works in order to take part in the joys of life. She lived without us and would have continued to live without us. A Butterfly's existence is not subject to ours, but rightfully independent of our aid.

Before the white-heart, the cauliflower, the savoy and the others were invented, the Pieris' caterpillar certainly did not lack food: he browsed on the wild cabbage of the cliffs, the parent of all the latter-day wealth; but, as this plant is not widely distributed and is, in any case, limited to certain maritime regions, the welfare of the Butterfly, whether on plain or hill, demanded a more luxuriant and more common plant for pasturage. This plant was apparently one of the Cruciferae, more or less seasoned with sulpheretted essence, like the cabbages. Let us experiment on these lines.

I rear the Pieris' caterpillars from the egg upwards on the wall-rocket (Diplotaxis tenuifolia, Dec.), which imbibes strong spices along the edge of the paths and at the foot of the walls. Penned in a large wire-gauze bell-cage, they accept this provender without demur; they nibble it with the same appetite as if it were cabbage; and they end by producing chrysalids and Butterflies. The change of fare causes not the least trouble.

I am equally successful with other crucifers of a less marked flavour: white mustard (Sinapis incana, Lin.), dyer's woad (Isatis tinctoria, Lin.), wild radish (Raphanus raphanistrum, Lin.), whitlow pepperwort (Lepidium draba, Lin.), hedge-mustard (Sisymbrium officinale, Scop.). On the other hand, the leaves of the lettuce, the bean, the pea, the corn-salad are obstinately refused. Let us be content with what we have seen: the fare has been sufficiently varied to show us that the cabbage-caterpillar feeds exclusively on a large number of crucifers, perhaps even on all.

As these experiments are made in the enclosure of a bell-cage, one might imagine that captivity impels the flock to feed, in the absence of better things, on what it would refuse were it free to hunt for itself. Having naught else within their reach, the starvelings consume any and all Cruciferae, without distinction of species. Can things sometimes be the same in the open fields, where I play none of my tricks? Can the family of the White Butterfly be settled on other Crucifers than the cabbage? I start a quest along the paths near the gardens and end by finding on wild radish and white mustard colonies as crowded and prosperous as those established on cabbage.

Now, except when the metamorphosis is at hand, the caterpillar of the White Butterfly never travels: he does all his growing on the identical plant whereon he saw the light. The caterpillars observed on the wild radish, as well as other households, are not, therefore, emigrants who have come as a matter of fancy from some cabbage-patch in the neighbourhood: they have hatched on the very leaves where I find them. Hence I arrive at this conclusion: the White Butterfly, who is fitful in her flight, chooses cabbage first, to dab her eggs upon, and different Cruciferae next, varying greatly in appearance.

How does the Pieris manage to know her way about her botanical domain? We have seen the Larini (A species of Weevils found on thistle-heads.—Translator's Note.), those explorers of fleshy receptacles with an artichoke flavour, astonish us with their knowledge of the flora of the thistle tribe; but their lore might, at a pinch, be explained by the method followed at the moment of housing the egg. With their rostrum, they prepare niches and dig out basins in the receptacle exploited and consequently they taste the thing a little before entrusting their eggs to it. On the other hand, the Butterfly, a nectar-drinker, makes not the least enquiry into the savoury qualities of the leafage; at most dipping her proboscis into the flowers, she abstracts a mouthful of syrup. This means of investigation, moreover, would be of no use to her, for the plant selected for the establishing of her family is, for the most part, not yet in flower. The mother flits for a moment around the plant; and that swift examination is enough: the emission of eggs takes place if the provender be found suitable.

The botanist, to recognize a crucifer, requires the indication provided by the flower. Here the Pieris surpasses us. She does not consult the seed-vessel, to see if it be long or short, nor yet the petals, four in number and arranged in a cross, because the plant, as a rule, is not in flower; and still she recognizes offhand what suits her caterpillars, in spite of profound differences that would embarrass any but a botanical expert.

Unless the Pieris has an innate power of discrimination to guide her, it is impossible to understand the great extent of her vegetable realm. She needs for her family Cruciferae, nothing but Cruciferae; and she knows this group of plants to perfection. I have been an enthusiastic botanist for half a century and more. Nevertheless, to discover if this or that plant, new to me, is or is not one of the Cruciferae, in the absence of flowers and fruits I should have more faith in the Butterfly's statements than in all the learned records of the books. Where science is apt to make mistakes instinct is infallible.

The Pieris has two families a year: one in April and May, the other in September. The cabbage-patches are renewed in those same months. The Butterfly's calendar tallies with the gardener's: the moment that provisions are in sight, consumers are forthcoming for the feast.

The eggs are a bright orange-yellow and do not lack prettiness when examined under the lens. They are blunted cones, ranged side by side on their round base and adorned with finely-scored longitudinal ridges. They are collected in slabs, sometimes on the upper surface, when the leaf that serves as a support is spread wide, sometimes on the lower surface when the leaf is pressed to the next ones. Their number varies considerably. Slabs of a couple of hundred are pretty frequent; isolated eggs, or eggs collected in small groups, are, on the contrary, rare. The mother's output is affected by the degree of quietness at the moment of laying.

The outer circumference of the group is irregularly formed, but the inside presents a certain order. The eggs are here arranged in straight rows backing against one another in such a way that each egg finds a double support in the preceding row. This alternation, without being of an irreproachable precision, gives a fairly stable equilibrium to the whole.

To see the mother at her laying is no easy matter: when examined too closely, the Pieris decamps at once. The structure of the work, however, reveals the order of the operations pretty clearly. The ovipositor swings slowly first in this direction, then in that, by turns; and a new egg is lodged in each space between two adjoining eggs in the previous row. The extent of the oscillation determines the length of the row, which is longer or shorter according to the layer's fancy.

The hatching takes place in about a week. It is almost simultaneous for the whole mass: as soon as one caterpillar comes out of its egg, the others come out also, as though the natal impulse were communicated from one to the other. In the same way, in the nest of the Praying Mantis, a warning seems to be spread abroad, arousing every one of the population. It is a wave propagated in all directions from the point first struck.

The egg does not open by means of a dehiscence similar to that of the vegetable-pods whose seeds have attained maturity; it is the new-born grub itself that contrives an exit-way by gnawing a hole in its enclosure. In this manner, it obtains near the top of the cone a symmetrical dormer-window, clean-edged, with no joins nor unevenness of any kind, showing that this part of the wall has been nibbled away and swallowed. But for this breach, which is just wide enough for the deliverance, the egg remains intact, standing firmly on its base. It is now that the lens is best able to take in its elegant structure. What it sees is a bag made of ultra-fine gold-beater's skin, translucent, stiff and white, retaining the complete form of the original egg. A score of streaked and knotted lines run from the top to the base. It is the wizard's pointed cap, the mitre with the grooves carved into jewelled chaplets. All said, the Cabbage-caterpillar's birth-casket is an exquisite work of art.

The hatching of the lot is finished in a couple of hours and the swarming family musters on the layer of swaddling-clothes, still in the same position. For a long time, before descending to the fostering leaf, it lingers on this kind of hot-bed, is even very busy there. Busy with what? It is browsing a strange kind of grass, the handsome mitres that remain standing on end. Slowly and methodically, from top to base, the new-born grubs nibble the wallets whence they have just emerged. By to-morrow, nothing is left of these but a pattern of round dots, the bases of the vanished sacks.

As his first mouthfuls, therefore, the Cabbage-caterpillar eats the membranous wrapper of his egg. This is a regulation diet, for I have never seen one of the little grubs allow itself to be tempted by the adjacent green stuff before finishing the ritual repast whereat skin bottles furnish forth the feast. It is the first time that I have seen a larva make a meal of the sack in which it was born. Of what use can this singular fare be to the budding caterpillar? I suspect as follows: the leaves of the cabbage are waxed and slippery surfaces and nearly always slant considerably. To graze on them without risking a fall, which would be fatal in earliest childhood, is hardly possible unless with moorings that afford a steady support. What is needed is bits of silk stretched along the road as fast as progress is made, something for the legs to grip, something to provide a good anchorage even when the grub is upside down. The silk-tubes, where those moorings are manufactured, must be very scantily supplied in a tiny, new-born animal; and it is expedient that they be filled without delay with the aid of a special form of nourishment. Then what shall the nature of the first food be? Vegetable matter, slow to elaborate and niggardly in its yield, does not fulfil the desired conditions at all well, for time presses and we must trust ourselves safely to the slippery leaf. An animal diet would be preferable: it is easier to digest and undergoes chemical changes in a shorter time. The wrapper of the egg is of a horny nature, as silk itself is. It will not take long to transform the one into the other. The grub therefore tackles the remains of its egg and turns it into silk to carry with it on its first journeys.

If my surmise is well-founded, there is reason to believe that, with a view to speedily filling the silk-glands to which they look to supply them with ropes, other caterpillars beginning their existence on smooth and steeply-slanting leaves also take as their first mouthful the membranous sack which is all that remains of the egg.

The whole of the platform of birth-sacks which was the first camping-ground of the White Butterfly's family is razed to the ground; naught remains but the round marks of the individual pieces that composed it. The structure of piles has disappeared; the prints left by the piles remain. The little caterpillars are now on the level of the leaf which shall henceforth feed them. They are a pale orange-yellow, with a sprinkling of white bristles. The head is a shiny black and remarkably powerful; it already gives signs of the coming gluttony. The little animal measures scarcely two millimetres in length. (.078 inch.—Translator's Note.)

The troop begins its steadying-work as soon as it comes into contact with its pasturage, the green cabbage-leaf. Here, there, in its immediate neighbourhood, each grub emits from its spinning glands short cables so slender that it takes an attentive lens to catch a glimpse of them. This is enough to ensure the equilibrium of the almost imponderable atom.

The vegetarian meal now begins. The grub's length promptly increases from two millimetres to four. Soon, a moult takes place which alters its costume: its skin becomes speckled, on a pale-yellow ground, with a number of black dots intermingled with white bristles. Three or four days of rest are necessary after the fatigue of breaking cover. When this is over, the hunger-fit starts that will make a ruin of the cabbage within a few weeks.

What an appetite! What a stomach, working continuously day and night! It is a devouring laboratory, through which the foodstuffs merely pass, transformed at once. I serve up to my caged herd a bunch of leaves picked from among the biggest: two hours later, nothing remains but the thick midribs; and even these are attacked when there is any delay in renewing the victuals. At this rate a "hundredweight-cabbage," doled out leaf by leaf, would not last my menagerie a week.

The gluttonous animal, therefore, when it swarms and multiplies, is a scourge. How are we to protect our gardens against it? In the days of Pliny, the great Latin naturalist, a stake was set up in the middle of the cabbage-bed to be preserved; and on this stake was fixed a Horse's skull bleached in the sun: a Mare's skull was considered even better. This sort of bogey was supposed to ward off the devouring brood.

My confidence in this preservative is but an indifferent one; my reason for mentioning it is that it reminds me of a custom still observed in our own days, at least in my part of the country. Nothing is so long-lived as absurdity. Tradition has retained in a simplified form, the ancient defensive apparatus of which Pliny speaks. For the Horse's skull our people have substituted an egg-shell on the top of a switch stuck among the cabbages. It is easier to arrange; also it is quite as useful, that is to say, it has no effect whatever.

Everything, even the nonsensical, is capable of explanation with a little credulity. When I question the peasants, our neighbours, they tell me that the effect of the egg-shell is as simple as can be: the Butterflies, attracted by the whiteness, come and lay their eggs upon it. Broiled by the sun and lacking all nourishment on that thankless support, the little caterpillars die; and that makes so many fewer.

I insist; I ask them if they have ever seen slabs of eggs or masses of young caterpillars on those white shells.

"Never," they reply, with one voice.

"Well, then?"

"It was done in the old days and so we go on doing it: that's all we know; and that's enough for us."

I leave it at that, persuaded that the memory of the Horse's skull, used once upon a time, is ineradicable, like all the rustic absurdities implanted by the ages.

We have, when all is said, but one means of protection, which is to watch and inspect the cabbage-leaves assiduously and crush the slabs of eggs between our finger and thumb and the caterpillars with our feet. Nothing is so effective as this method, which makes great demands on one's time and vigilance. What pains to obtain an unspoilt cabbage! And what a debt do we not owe to those humble scrapers of the soil, those ragged heroes, who provide us with the wherewithal to live!

To eat and digest, to accumulate reserves whence the Butterfly will issue: that is the caterpillar's one and only business. The Cabbage-caterpillar performs it with insatiable gluttony. Incessantly it browses, incessantly digests: the supreme felicity of an animal which is little more than an intestine. There is never a distraction, unless it be certain see-saw movements which are particularly curious when several caterpillars are grazing side by side, abreast. Then, at intervals, all the heads in the row are briskly lifted and as briskly lowered, time after time, with an automatic precision worthy of a Prussian drill-ground. Can it be their method of intimidating an always possible aggressor? Can it be a manifestation of gaiety, when the wanton sun warms their full paunches? Whether sign of fear or sign of bliss, this is the only exercise that the gluttons allow themselves until the proper degree of plumpness is attained.

After a month's grazing, the voracious appetite of my caged herd is assuaged. The caterpillars climb the trelliswork in every direction, walk about anyhow, with their forepart raised and searching space. Here and there, as they pass, the swaying herd put forth a thread. They wander restlessly, anxiously to travel afar. The exodus now prevented by the trellised enclosure I once saw under excellent conditions. At the advent of the cold weather, I had placed a few cabbage-stalks, covered with caterpillars, in a small greenhouse. Those who saw the common kitchen vegetable sumptuously lodged under glass, in the company of the pelargonium and the Chinese primrose, were astonished at my curious fancy. I let them smile. I had my plans: I wanted to find out how the family of the Large White Butterfly behaves when the cold weather sets in. Things happened just as I wished. At the end of November, the caterpillars, having grown to the desired extent, left the cabbages, one by one, and began to roam about the walls. None of them fixed himself there or made preparations for the transformation. I suspected that they wanted the choice of a spot in the open air, exposed to all the rigours of winter. I therefore left the door of the hothouse open. Soon the whole crowd had disappeared.

I found them dispersed all over the neighbouring walls, some thirty yards off. The thrust of a ledge, the eaves formed by a projecting bit of mortar served them as a shelter where the chrysalid moult took place and where the winter was passed. The Cabbage-caterpillar possesses a robust constitution, unsusceptible to torrid heat or icy cold. All that he needs for his metamorphosis is an airy lodging, free from permanent damp.

The inmates of my fold, therefore, move about for a few days on the trelliswork, anxious to travel afar in search of a wall. Finding none and realizing that time presses, they resign themselves. Each one, supporting himself on the trellis, first weaves around himself a thin carpet of white silk, which will form the sustaining layer at the time of the laborious and delicate work of the nymphosis. He fixes his rear-end to this base by a silk pad and his fore-part by a strap that passes under his shoulders and is fixed on either side to the carpet. Thus slung from his three fastenings, he strips himself of his larval apparel and turns into a chrysalis in the open air, with no protection save that of the wall, which the caterpillar would certainly have found had I not interfered.

Of a surety, he would be short-sighted indeed that pictured a world of good things prepared exclusively for our advantage. The earth, the great foster-mother, has a generous breast. At the very moment when nourishing matter is created, even though it be with our own zealous aid, she summons to the feast host upon host of consumers, who are all the more numerous and enterprising in proportion as the table is more amply spread. The cherry of our orchards is excellent eating: a maggot contends with us for its possession. In vain do we weigh suns and planets: our supremacy, which fathoms the universe, cannot prevent a wretched worm from levying its toll on the delicious fruit. We make ourselves at home in a cabbage bed: the sons of the Pieris make themselves at home there too. Preferring broccoli to wild radish, they profit where we have profited; and we have no remedy against their competition save caterpillar-raids and egg-crushing, a thankless, tedious, and none too efficacious work.

Every creature has its claims on life. The Cabbage-caterpillar eagerly puts forth his own, so much so that the cultivation of the precious plant would be endangered if others concerned did not take part in its defence. These others are the auxiliaries (The author employs this word to denote the insects that are helpful, while describing as "ravagers" the insects that are hurtful to the farmer's crops.— Translator's Note.), our helpers from necessity and not from sympathy. The words friend and foe, auxiliaries and ravagers are here the mere conventions of a language not always adapted to render the exact truth. He is our foe who eats or attacks our crops; our friend is he who feeds upon our foes. Everything is reduced to a frenzied contest of appetites.

In the name of the might that is mine, of trickery, of highway robbery, clear out of that, you, and make room for me: give me your seat at the banquet! That is the inexorable law in the world of animals and more or less, alas, in our own world as well!

Now, among our entomological auxiliaries, the smallest in size are the best at their work. One of them is charged with watching over the cabbages. She is so small, she works so discreetly that the gardener does not know her, has not even heard of her. Were he to see her by accident, flitting around the plant which she protects, he would take no notice of her, would not suspect the service rendered. I propose to set forth the tiny midget's deserts.

Scientists call her Microgaster glomeratus. What exactly was in the mind of the author of the name Microgaster, which means little belly? Did he intend to allude to the insignificance of the abdomen? Not so. However slight the belly may be, the insect nevertheless possesses one, correctly proportioned to the rest of the body, so that the classic denomination, far from giving us any information, might mislead us, were we to trust it wholly. Nomenclature, which changes from day to day and becomes more and more cacophonous, is an unsafe guide. Instead of asking the animal what its name is, let us begin by asking:

"What can you do? What is your business?"

Well, the Microgaster's business is to exploit the Cabbage-caterpillar, a clearly-defined business, admitting of no possible confusion. Would we behold her works? In the spring, let us inspect the neighbourhood of the kitchen-garden. Be our eye never so unobservant, we shall notice against the walls or on the withered grasses at the foot of the hedges some very small yellow cocoons, heaped into masses the size of a hazel-nut.

Beside each group lies a Cabbage-caterpillar, sometimes dying, sometimes dead, and always presenting a most tattered appearance. These cocoons are the work of the Microgaster's family, hatched or on the point of hatching into the perfect stage; the caterpillar is the dish whereon that family has fed during its larval state. The epithet glomeratus, which accompanies the name of Microgaster, suggests this conglomeration of cocoons. Let us collect the clusters as they are, without seeking to separate them, an operation which would demand both patience and dexterity, for the cocoons are closely united by the inextricable tangle of their sur-face-threads. In May a swarm of pigmies will sally forth, ready to get to business in the cabbages.

Colloquial language uses the terms Midge and Gnat to describe the tiny insects which we often see dancing in a ray of sunlight. There is something of everything in those aerial ballets. It is possible that the persecutrix of the Cabbage-caterpillar is there, along with many another; but the name of Midge cannot properly be applied to her. He who says Midge says Fly, Dipteron, two-winged insect; and our friend has four wings, one and all adapted for flying. By virtue of this character-istic and others no less important, she belongs to the order of Hymenoptera. (This order includes the Ichneumon-flies, of whom the Microgaster is one.—Translator's Note.) No matter: as our language possesses no more precise term outside the scientific vocabulary, let us use the expression Midge, which pretty well conveys the general idea. Our Midge, the Microgaster, is the size of an aver-age Gnat. She measures 3 or 4 millimetres. (.117 to .156 inch.—Translator's Note.) The two sexes are equally numerous and wear the same costume, a black uniform, all but the legs, which are pale red. In spite of this likeness, they are eas-ily distinguished. The male has an abdomen which is slightly flattened and, more-over, curved at the tip; the female, before the laying, has hers full and perceptibly distended by its ovular contents. This rapid sketch of the insect should be enough for our purpose.

If we wish to know the grub and especially to inform ourselves of its manner of living, it is advisable to rear in a cage a numerous herd of Cabbage-caterpillars. Whereas a direct search on the cabbages in our garden would give us but a diffi-cult and uncertain harvest, by this means we shall daily have as many as we wish before our eyes.

In the course of June, which is the time when the caterpillars quit their pastures and go far afield to settle on some wall or other, those in my fold, finding nothing better, climb to the dome of the cage to make their preparations and to spin a supporting network for the chrysalid's needs. Among these spinners we see some weaklings working listlessly at their carpet. Their appearance makes us deem them in the grip of a mortal disease. I take a few of them and open their bellies, using a needle by way of a scalpel. What comes out is a bunch of green entrails, soaked in a bright yellow fluid, which is really the creature's blood. These tangled intestines swarm with little lazy grubs, varying greatly in number, from ten or twenty at least to sometimes half a hundred. They are the offspring of the Microgaster.

What do they feed on? The lens makes conscientious enquiries; nowhere does it manage to show me the vermin attacking solid nourishment, fatty tissues, muscles or other parts; nowhere do I see them bite, gnaw, or dissect. The following experiment will tell us more fully: I pour into a watch-glass the crowds extracted from the hospitable paunches. I flood them with caterpillar's blood obtained by simple pricks; I place the preparation under a glass bell-jar, in a moist atmosphere, to prevent evaporation; I repeat the nourishing bath by means of fresh bleedings and give them the stimulant which they would have gained from the living caterpillar. Thanks to these precautions, my charges have all the appearance of excellent health; they drink and thrive. But this state of things cannot last long. Soon ripe for the transformation, my grubs leave the dining-room of the watch-glass as they would have left the caterpillar's belly; they come to the ground to try and weave their tiny cocoons. They fail in the attempt and perish. They have missed a suitable support, that is to say, the silky carpet provided by the dying caterpillar. No matter: I have seen enough to convince me. The larvae of the Microgaster do not eat in the strict sense of the word; they live on soup; and that soup is the caterpillar's blood.

Examine the parasites closely and you shall see that their diet is bound to be a liquid one. They are little white grubs, neatly segmented, with a pointed forepart splashed with tiny black marks, as though the atom had been slaking its thirst in a drop of ink. It moves its hind-quarters slowly, without shifting its position. I place it under the microscope. The mouth is a pore, devoid of any apparatus for disintegration-work: it has no fangs, no horny nippers, no mandibles; its attack is just a kiss. It does not chew, it sucks, it takes discreet sips at the moisture all around it.

The fact that it refrains entirely from biting is confirmed by my autopsy of the stricken caterpillars. In the patient's belly, notwithstanding the number of nurselings who hardly leave room for the nurse's entrails, everything is in perfect order; nowhere do we see a trace of mutilation. Nor does aught on the outside betray any havoc within. The exploited caterpillars graze and move about peacefully, giving no sign of pain. It is impossible for me to distinguish them from the unscathed ones in respect of appetite and untroubled digestion.

When the time approaches to weave the carpet for the support of the chrysalis, an appearance of emaciation at last points to the evil that is at their vitals. They spin nevertheless. They are stoics who do not forget their duty in the hour of death. At last they expire, quite softly, not of any wounds, but of anaemia, even as a lamp goes out when the oil comes to an end. And it has to be. The living caterpillar, capable of feeding himself and forming blood, is a necessity for the welfare of the grubs; he has to last about a month, until the Microgaster's offspring have achieved their full growth. The two calendars synchronize in a remarkable way. When the caterpillar leaves off eating and makes his preparations for the metamorphosis, the parasites are ripe for the exodus. The bottle dries up when the drinkers cease to need it; but until that moment it must remain more or less well-filled, although becoming limper daily. It is important, therefore, that the caterpillar's existence be not endangered by wounds which, even though very tiny, would stop the working of the blood-fountains. With this intent, the drainers of the bottle are, in a manner of speaking, muzzled; they have by way of a mouth a pore that sucks without bruising.

The dying caterpillar continues to lay the silk of his carpet with a slow oscillation of the head. The moment now comes for the parasites to emerge. This happens in June and generally at nightfall. A breach is made on the ventral surface or else in the sides, never on the back: one breach only, contrived at a point of minor resistance, at the junction of two segments; for it is bound to be a toilsome business, in the absence of a set of filing-tools. Perhaps the grubs take one another's places at the point attacked and come by turns to work at it with a kiss.

In one short spell, the whole tribe issues through this single opening and is soon wriggling about, perched on the surface of the caterpillar. The lens cannot perceive the hole, which closes on the instant. There is not even a haemorrhage: the bottle has been drained too thoroughly. You must press it between your fingers to squeeze out a few drops of moisture and thus discover the place of exit.

Around the caterpillar, who is not always quite dead and who sometimes even goes on weaving his carpet a moment longer, the vermin at once begin to work at their cocoons. The straw-coloured thread, drawn from the silk-glands by a backward jerk of the head, is first fixed to the white network of the caterpillar and then produces adjacent warp-beams, so that, by mutual entanglements, the individual works are welded together and form an agglomeration in which each of the grubs has its own cabin. For the moment, what is woven is not the real cocoon, but a general scaffolding which will facilitate the construction of the separate shells. All these frames rest upon those adjoining and, mixing up their threads, become a common edifice wherein each grub contrives a shelter for itself. Here at last the real cocoon is spun, a pretty little piece of closely-woven work.

In my rearing-jars I obtain as many groups of these tiny shells as my future experiments can wish for. Three-fourths of the caterpillars have supplied me with them, so ruthless has been the toll of the spring births. I lodge these groups, one by one, in separate glass tubes, thus forming a collection on which I can draw at will, while, in view of my experiments, I keep under observation the whole swarm produced by one caterpillar.

The adult Microgaster appears a fortnight later, in the middle of June. There are fifty in the first tube examined. The riotous multitude is in the full enjoyment of the pairing-season, for the two sexes always figure among the guests of any one caterpillar. What animation! What an orgy of love! The carnival of these pigmies bewilders the observer and makes his head swim.

Most of the females, wishful of liberty, plunge down to the waist between the glass of the tube and the plug of cotton-wool that closes the end turned to the light; but the lower halves remain free and form a circular gallery in front of which the males hustle one another, take one another's places and hastily operate. Each bides his turn, each attends to his little matters for a few moments and then makes way for his rivals and goes off to start again elsewhere. The turbulent wedding lasts all the morning and begins afresh next day, a mighty throng of couples embracing, separating and embracing once more.

There is every reason to believe that, in gardens, the mated ones, finding themselves in isolated couples, would keep quieter. Here, in the tube, things degenerate into a riot because the assembly is too numerous for the narrow space.

What is lacking to complete its happiness? Apparently a little food, a few sugary mouthfuls extracted from the flowers. I serve up some provisions in the tubes: not drops of honey, in which the puny creatures would get stuck, but little strips of paper spread with that dainty. They come to them, take their stand on them and refresh themselves. The fare appears to agree with them. With this diet, renewed as the strips dry up, I can keep them in very good condition until the end of my inquisition.

There is another arrangement to be made. The colonists in my spare tubes are restless and quick of flight; they will have to be transferred presently to sundry vessels without my risking the loss of a good number, or even the whole lot, a loss which my hands, my forceps and other means of coercion would be unable to prevent by checking the nimble movements of the tiny prisoners. The irresistible attraction of the sunlight comes to my aid. If I lay one of my tubes horizontally on the table, turning one end towards the full light of a sunny window, the captives at once make for the brighter end and play about there for a long while, without seeking to retreat. If I turn the tube in the opposite direction, the crowd immediately shifts its quarters and collects at the other end. The brilliant sunlight is its great joy. With this bait, I can send it whithersoever I please.

We will therefore place the new receptacle, jar or test-tube, on the table, pointing the closed end towards the window. At its mouth, we open one of the full tubes. No other precaution is needed: even though the mouth leaves a large interval free, the swarm hastens into the lighted chamber. All that remains to be done is to close the apparatus before moving it. The observer is now in control of the multitude, without appreciable losses, and is able to question it at will.

We will begin by asking:

"How do you manage to lodge your germs inside the caterpillar?"

This question and others of the same category, which ought to take precedence of everything else, are generally neglected by the impaler of insects, who cares more for the niceties of nomenclature than for glorious realities. He classifies his subjects, dividing them into regiments with barbarous labels, a work which seems to him the highest expression of entomological science. Names, nothing but names: the rest hardly counts. The persecutor of the Pieris used to be called Microgaster, that is to say, little belly: to-day she is called Apanteles, that is to say, the incomplete. What a fine step forward! We now know all about it!

Can our friend at least tell us how "the Little Belly" or "the Incomplete" gets into the caterpillar? Not a bit of it! A book which, judging by its recent date, should be the faithful echo of our actual knowledge, informs us that the Microgaster inserts her eggs direct into the caterpillar's body. It goes on to say that the parasitic vermin inhabit the chrysalis, whence they make their way out by perforating the stout horny wrapper. Hundreds of times have I witnessed the exodus of the grubs ripe for weaving their cocoons; and the exit has always been made through the skin of the caterpillar and never through the armour of the chrysalis. The fact that its mouth is a mere clinging pore, deprived of any offensive weapon, would even lead me to believe that the grub is incapable of perforating the chrysalid's covering.

This proved error makes me doubt the other proposition, though logical, after all, and agreeing with the methods followed by a host of parasites. No matter: my faith in what I read in print is of the slightest; I prefer to go straight to facts. Before making a statement of any kind, I want to see, what I call seeing. It is a slower and more laborious process; but it is certainly much safer.

I will not undertake to lie in wait for what takes place on the cabbages in the garden: that method is too uncertain and besides does not lend itself to precise observation. As I have in hand the necessary materials, to wit, my collection of tubes swarming with the parasites newly hatched into the adult form, I will operate on the little table in my animals' laboratory. A jar with a capacity of about a litre (About 1 3/4 pints, or .22 gallon.—Translator's Note.) is placed on the table, with the bottom turned towards the window in the sun. I put into it a cabbage-leaf covered with caterpillars, sometimes fully developed, sometimes half-way, sometimes just out of the egg. A strip of honeyed paper will serve the Microgaster as a dining room, if the experiment is destined to take some time. Lastly, by the method of transfer which I described above, I send the inmates of one of my tubes into the apparatus. Once the jar is closed, there is nothing left to do but to let things take their course and to keep an assiduous watch, for days and weeks, if need be. Nothing worth remarking can escape me.

The caterpillars graze placidly, heedless of their terrible attendants. If some giddy-pates in the turbulent swarm pass over the caterpillars' spines, these draw up their fore-part with a jerk and as suddenly lower it again; and that is all: the intruders forthwith decamp. Nor do the latter seem to contemplate any harm: they refresh themselves on the honey-smeared strip, they come and go tumultuously. Their short flights may land them, now in one place, now in another, on the browsing herd, but they pay no attention to it. What we see is casual meetings, not deliberate encounters.

In vain I change the flock of caterpillars and vary their age; in vain I change the squad of parasites; in vain I follow events in the jar for long hours, morning and evening, both in a dim light and in the full glare of the sun: I succeed in seeing nothing, absolutely nothing, on the parasite's side, that resembles an attack. No matter what the ill-informed authors say—ill-informed because they had not the patience to see for themselves—the conclusion at which I arrive is positive: to inject the germs, the Microgaster never attacks the caterpillars.

The invasion, therefore, is necessarily effected through the Butterfly's eggs themselves, as experiment will prove. My broad jar would tell against the inspection of the troop, kept at too great a distance by the glass enclosure, and I therefore select a tube an inch wide. I place in this a shred of cabbage-leaf, bearing a slab of eggs, as laid by the Butterfly. I next introduce the inmates of one of my spare vessels. A strip of paper smeared with honey accompanies the new arrivals.

This happens early in July. Soon, the females are there, fussing about, sometimes to the extent of blackening the whole slab of yellow eggs. They inspect the treasure, flutter their wings and brush their hind-legs against each other, a sign of keen satisfaction. They sound the heap, probe the interstices with their antennae and tap the individual eggs with their palpi; then, this one here, that one there, they quickly apply the tip of their abdomen to the egg selected. Each time, we see a slender, horny prickle darting from the ventral surface, close to the end. This is the instrument that deposits the germ under the film of the egg; it is the inoculation-needle. The operation is performed calmly and methodically, even when several mothers are working at one and the same time. Where one has been, a second goes, followed by a third, a fourth and others yet, nor am I able definitely to see the end of the visits paid to the same egg. Each time, the needle enters and inserts a germ.

It is impossible, in such a crowd, for the eye to follow the successive mothers who hasten to lay in each; but there is one quite practicable method by which we can estimate the number of germs introduced into a single egg, which is, later, to open the ravaged caterpillars and count the grubs which they contain. A less repugnant means is to number the little cocoons heaped up around each dead caterpillar. The total will tell us how many germs were injected, some by the same mother returning several times to the egg already treated, others by different mothers. Well, the number of these cocoons varies greatly. Generally, it fluctuates in the neighbourhood of twenty, but I have come across as many as sixty-five; and nothing tells me that this is the extreme limit. What hideous industry for the extermination of a Butterfly's progeny!

I am fortunate at this moment in having a highly-cultured visitor, versed in the profundities of philosophic thought. I make way for him before the apparatus wherein the Microgaster is at work. For an hour and more, standing lens in hand, he, in his turn, looks and sees what I have just seen; he watches the layers who go from one egg to the other, make their choice, draw their slender lancet and prick what the stream of passers-by, one after the other, have already pricked. Thoughtful and a little uneasy, he puts down his lens at last. Never had he been vouchsafed so clear a glimpse as here, in my finger-wide tube, of the masterly brigandage that runs through all life down to that of the very smallest.

INDEX.

Blackbirds, Corsican.

Bluebottle.
the laying of the eggs.
hatching.
a test.
paper a protection against.
the grubs.
sand a protection against.

Bower-bird.

Brussels Sprouts, ancestry of.

Buprestis.

Burying-beetles: method of burial.
appearance of the insect.
manipulation of the corpse.
cooperation of individuals.
larvae of.
attacked by vermin.
the dismal end of.
experiments.
test conditions imposed.
conditions of burial.
nets of cordage cut through.
ligatures severed.
limitations of instinct.

Cabbage, ancestry of.
offspring.

Cabbage Butterfly, her selection of suitable Cruciferae.
eggs of.
hatching of the eggs.

Cabbage-caterpillar.
eats egg-cases on emergence.
employment of silk by.

growth and moults.
its voracity.
an old charm against.
the only true charm.
movements of the caterpillar.
its chrysalis.
its deadly enemy.

Calliphora vomitaria, see Bluebottle.

Capricorn Beetle.
the grub.
its cell.
the barricade.
the pupa.
metamorphosis and emergence.

Cauliflower.

Centauries.

Cerambyx miles.

Cerceris.

Cetonia, or Rose-chafer.

Chalicodoma.

Chat, Black-eared.

Cicada.
the grasshopper's victim.

Cicadella.

Clairville on the Burying-beetle.

Clothes-moth.

Cockchafers.

Snail, the prey of the Glow-worm.

Sphex.

Sphex, White-banded.

Spiders.
apprised of prey by vibration.

Staphylinus.

Stizus.

Swede.

Tadpoles.

Tarantula, Black-bellied, see Lycosa.

Thistles.

Thomisus.

Toad, Bell-ringing.

Tree-frogs.

Tree Wasps.

Turkeys, how trapped.

Ventoux, Mount.

Wasp, Common.

Woodpecker.

Printed in the United Kingdom
by Lightning Source UK Ltd.
119763UK00003B/125